THE BIOLOGY OF RARITY

Population and Community Biology Series

Principal Editor

M.B. Usher
Chief Scientific Adviser and Director of Research and Advisory Services, Scottish Natural Heritage, UK

Editors

D.L. DeAngelis
Senior Scientist, Environmental Sciences Division, Oak Ridge National Laboratory, USA

B.F.J. Manly
Director, Centre for Application of Statistics and Mathematics, University of Otago, New Zealand

The study of both populations and communities is central to the science of ecology. This series of books explores many facets of population biology and the processes that determine the structure and dynamics of communities. Although individual authors are given freedom to develop their subjects in their own way, these books are scientifically rigorous and a quantitative approach to analysing population and community phenomena is often used.

Already published

7. **Stage-Structured Populations: Sampling, Analysis and Simulation**
   B.F.J. Manly (1990) 200pp. Hb.

8. **Habitat Structure: The Physical Arrangement of Objects in Space**
   S.S. Bell, E.D. McCoy and H.R. Mushinsky (1991) 452pp. Hb.

9. **Dynamics of Nutrient Cycling and Food Webs**
   D.L. DeAngelis (1992) 285pp. Pb.

10. **Analytical Population Dynamics**
    T. Royama (1992) 387pp. Hb/Pb.

11. **Plant Succession: Theory and Prediction**
    D.C. Glenn-Lewin, R.K. Peet and T.T. Veblen (1992) 361pp. Hb.

12. **Risk Assessment in Conservation Biology**
    M.A. Burgman, S. Ferson and R. Akcakaya (1993) 324pp. Hb.

13. **Rarity**
    K. Gaston (1994) 192pp. Hb/Pb.

14. **Fire and Plants**
    W.J. Bond and B.W. van Wilgen (1996) 288pp. Hb.

15. **Biological Invasions**
    M. Williamson (1996) 256pp. Pb.

16. **Regulation and Stabilization: Paradigms in Population Biology**
    P.J. den Boer and J. Reddingius (1996) 400pp. Hb.

THE BIOLOGY OF RARITY
Causes and consequences of rare–common differences

Edited by

William E. Kunin

Lecturer in Ecology, Department of Biology, University of Leeds, Leeds, UK

and

Kevin J. Gaston

Royal Society University Research Fellow, Department of Animal and Plant Sciences, University of Sheffield, Sheffield, UK

CHAPMAN & HALL

London · Weinheim · New York · Tokyo · Melbourne · Madras

Published by Chapman & Hall, 2–6 Boundary Row, London SE1 8HN, UK

Chapman & Hall, 2–6 Boundary Row, London SE1 8HN, UK

Chapman & Hall GmbH, Pappelallee 3, 69469 Weinheim, Germany

Chapman & Hall USA, 115 Fifth Avenue, New York NY 10003, USA

Chapman & Hall Japan, ITP-Japan, Kyowa Building, 3F, 2-2-1 Hirakawacho, Chiyoda-ku, Tokyo 102, Japan

Chapman & Hall Australia, 102 Dodds Street, South Melbourne, Victoria 3205, Australia

Chapman & Hall India, R. Seshadri, 32 Second Main Road, CIT East, Madras 600 035, India

First edition 1997

© 1997 Chapman and Hall

Typeset in 10/12pt Times by Cambrian Typesetters, Frimley, Surrey

Printed in Great Britain by St Edmundsbury Press, Suffolk

ISBN 0 412 63380 9

A catalogue record for this book is available from the British Library

Library of Congress Catalog Card Number: 96–71014

∞ Printed on permanent acid-free text paper, manufactured in accordance with ANSI/NISO Z39.48-1992 and ANSI/NISO Z39.48-1984 (Permanence of Paper).

Contents

Contributors

Tim M. Blackburn
Centre for Population Biology
Imperial College at Silwood Park
Ascot, Berkshire SL5 7PY
UK

Steven L. Chown
Department of Zoology and Entomology
University of Pretoria
Zoology Building, Room 3–7
Pretoria 0002
South Africa

Peter Cotgreave
Institute of Zoology
Zoological Society of London
Regent's Park
London NW1 4RY
UK

Kevin J. Gaston
Department of Animal and Plant Sciences
University of Sheffield
Sheffield S10 2TN
UK

Robert D. Holt
Museum of Natural History
Department of Systematics and Ecology
University of Kansas
Lawrence
Kansas 66045
USA

Jeffrey D. Karron
Department of Biological Sciences
University of Wisconsin, Milwaukee
Lapham Hall
PO Box 413
Milwaukee
Wisconsin 53201
USA

Melanie Kershaw
Institute of Zoology
Zoological Society of London
Regent's Park
London NW1 4RY
UK

William E. Kunin
Department of Biology
University of Leeds
Leeds LS2 9JT
UK

Mark V. Lomolino
Department of Biology and Oklahoma Biological Survey
University of Oklahoma
Norman
Oklahoma 73019
USA

Georgina M. Mace
Institute of Zoology
Zoological Society of London
Regent's Park
London NW1 4RY
UK

Michael L. McKinney
Department of Geological Sciences
306 G&G Building
University of Tennessee
Knoxville
Tennessee 37996-1410
USA

Gordon H. Orians
Department of Zoology
Box 351800
University of Washington
Seattle
Washington 98195
USA

Mark D. Pagel
Department of Zoology
South Parks Road
University of Oxford
Oxford OX1 3PS
UK

Michael L. Rosenzweig
Department of Ecology and Evolutionary Biology
University of Arizona
Tucson
Arizona 85721
USA

Preface

This book began life as a review article. That article spawned a symposium which was, in turn, greatly expanded to form the present volume. As the project moved through these developmental stages (hopefully, towards attainment of its full maturity), a number of people have provided invaluable assistance to us, and we would like to take this opportunity to thank them.

Gordon Orians must certainly take a high place in that list. He has been both a friend and mentor to W.E.K., and many of the topics explored in this book have emerged from the resultant dialogue. His thought processes, ideas and perhaps even some of his turns of phrase emerge throughout much of the book. Gordon also played a pivotal role in inviting the article that set this project in motion, and so he has served as a catalyst to the book as well as one of its reagents. While he has not served as an editor of this book, he is one of its authors in more than just the literal sense.

John Lawton has likewise been a key figure behind the scenes in this venture, and even though his name does not appear on any of this book's chapters, his signature is to be found everywhere within it. He has been an adviser, collaborator and colleague to both of us. He and Tim Blackburn brought the two of us together initially, and he provided the physical and intellectual setting, in the NERC Centre for Population Biology at Silwood Park, where that interaction could flourish.

Numerous other colleagues have facilitated the development of our thoughts on rarity, in many discussions both verbal and written. In particular, we are grateful to Tim Blackburn, Avi Shmida, Ian Gauld, Dan Cohen, Phil Warren, Moshe Shachak and Paul Williams.

In each of the stages of this project, we have been the beneficiaries of capable administration and expert advice from a variety of sources. We would especially like to thank Andrew Sugden of *Trends in Ecology and Evolution*, Dennis Whigham and the Program Committee of the Ecological Society of America, and Bob Carling, Helen Sharples, Nicola Denny, Paul Gill and Mike Usher of Chapman & Hall. Collectively they have shepherded us and this book over countless obstacles, and seen us and it safely to an end that we never anticipated at the outset.

We would also like to thank those who have served as referees for the chapters in this volume. Every chapter has been through a peer review process, and many have been improved markedly as a result. We owe a great debt to the colleagues and friends who succumbed to the combination of curiosity, enthusiasm, guilt and desperation that we brought to bear. To preserve their anonymity, we cannot name most of the reviewers, but two among them forfeited their privacy by signing their comments. We therefore thank Christophe Thebaud, John Lawton and the other 15 busy people who have contributed their time to improving the quality of this book.

We also asked many of the chapter authors to take part in some mutual reviewing, in the hope of creating an internal dialogue that might serve to bring together some of the very different strands contained in this volume. We would like to thank them for their creativity, their patience and their commitment to this book; and hope they will find the finished product worthy of the effort. They and all of those named above can take a measure of the credit for whatever is good, clear and interesting in this book. Whatever faults remain are, of course, our own responsibility.

Bill Kunin
Leeds

Kevin J. Gaston
Sheffield

PART ONE: RARITY AND RARE–COMMON DIFFERENCES

1 Introduction: on the causes and consequences of rare–common differences

William E. Kunin

1.1 INTRODUCTION

The purpose of this volume is to begin the complicated process of separating pattern from process. In a sense, it is an exercise that could be performed with any interesting pattern of species and their associated characteristics. We choose to focus our attention here on one particularly intriguing case: the differences between rare and common species. The choice is not completely arbitrary. The causes and consequences of species abundance have been a central concern of ecologists since the discipline began; indeed, some define ecology as the study of the distribution and abundance of species. The rise of conservation biology as a sub-discipline in recent years has served to focus additional attention on rarity and its implications, and has added to the importance of understanding the nature of rare–common differences.

If rarity is a uniquely intriguing subject, it is also a uniquely problematic one. Rarity is a species characteristic, but not in the same sense that hair colour or wing venation or other morphological traits are; it is an emergent trait of a species' population and its environment rather than a trait of an individual organism. This has a number of important consequences. Abundance, as a population trait, can change much more quickly than a morphological trait can evolve, and yet such changes are much less likely to be preserved in the fossil record (Chapter 7). Similarly, when speciation occurs, the rarity or abundance of a parent species may not be directly inherited by its daughter taxa in the usual genetic sense of the word (although abundance may be linked to traits which are heritable; Jablonski, 1987). Perhaps most significantly, the emergent nature of a species' abundance makes it an unusually difficult characteristic to observe, especially in poorly studied groups – you cannot record a

The Biology of Rarity. Edited by William E. Kunin and Kevin J. Gaston.
Published in 1997 by Chapman & Hall, London. ISBN 0 412 63380 9.

species' abundance from a single museum specimen. Even where information is plentiful, it is not always clear how abundance should be measured; it is a complex phenomenon, fraught with methodological and terminological difficulties (Chapter 3; Kunin and Gaston, 1993; Gaston, 1994). All of these points raise important issues that will haunt and enliven most of the chapters that follow. They may make our task more difficult, but they also make it more interesting.

If we are to examine the relationship between pattern and process, we must start by establishing the pattern. There is a tendency, reported repeatedly (but less than universally) in the literature, for rare and common species to exhibit somewhat different suites of characteristics (reviewed in Chapter 2). The differences are often subtle and are statistical rather than absolute, and there are almost always counter-examples to any published trend. Nonetheless, despite all the noise and methodological complications, there is growing evidence that rare and common species (however defined) differ overall in a number of important respects. Such a conclusion only leads to more questions. Where rare species differ from common, how is it that these differences developed? What mechanisms are responsible for the observed pattern?

The possibilities are much more numerous than it would appear at first glance. For convenience, they will be divided into three functional groups: entry rules, exit rules and transformations (Figure 1.1); that is: biases in the processes that determine who joins the set of rare species; biases in the processes that determine who leaves it; and changes caused by processes that act upon species while they are members of that set. These categories can perhaps be best explained by way of an analogy (D. Kunin, personal communication). If, upon arriving at a party that had been under way for some time, a visitor noticed that most of the people in attendance were drunk, there would be a number of possible hypotheses he or she might employ to explain the pattern. The first, of course, is the null hypothesis: that the party-goers represent a random collection of people (with their drunkenness either reflecting a general inebriation of the local populace or being the result of a statistical fluke). If, on the other hand, the preponderance of drunkenness at the party is too high relative to that of the surrounding community to be explained by chance alone, it may be that the invitations to the party (the entry rules) were biased; the host may have a particular fondness for people with drinking problems. Alternatively, the exit rules for the party may be biased: everyone sober enough to drive may have left already. Finally, there may be some transforming activity happening at the party (it is left to the reader's imagination) that has changed formerly sober citizens into drunkards.

In a similar way, the set of rare species can become biased by the laws that govern who becomes rare in the first place; it can be influenced by the processes (primarily, extinction) that determine which of those species leave the set; and/or the conditions of rarity can set in motion processes

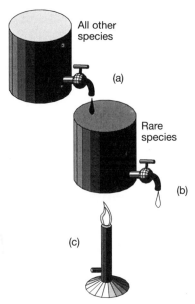

Figure 1.1 Difference between entry rules, exit rules and transformations. The set of rare species can be affected by (a) biases in the input of species from the general pool, (b) biases in the departure of species from the set, or (c) processes that transform them while they are in it.

that cause the properties of species to change. Each of these possibilities can be further elaborated into a number of competing (but not necessarily mutually exclusive) scenarios. The next few paragraphs give a rough sketch of a few of these possibilities. Most of the rest of this book will be devoted to exploring them in greater detail.

1.2 A CASE IN POINT: BREEDING SYSTEMS AND ABUNDANCE

To make the subject a bit less abstract, let us concentrate here on one particular pattern: the tendency of rare plant species to use asexuality or self-fertilization in reproduction. A number of authors (e.g. Hamilton, 1990; Kunin, 1991; Longton, 1992; but not Karron, 1987) have noted that rare species tend to be self-compatible more often than one would expect by chance, if they were a random subsample of all plants. The difference is certainly not absolute; there are many self-incompatible rare species, and quite a few self-compatible common ones. Nonetheless, the pattern seems to hold up as a broad generalization. What is responsible for this trend? There are possible explanations that fit under each of the three broad categories discussed above.

1.2.1 Entry rules

(a) Selfing as an evolutionary dead end – Müller's ratchet and related pitfalls

It seems unlikely that species abundances are assigned at random. For example, a species that is poorly adapted to its environment, or one adapted to only a very limited range of conditions, seems likely to become rare – perhaps even as a prelude to its extinction. Asexuality, it has often been suggested, is likely to lead to such difficulties. An asexually reproducing population cannot easily adapt to changing conditions. Worse yet, it should gradually accumulate deleterious mutations (via 'Müller's ratchet': Müller, 1964) leading to a further deterioration of its performance. In a 'Red Queen' world where everyone else must run fast just to hold still, asexual species may actually be tiptoeing backwards. Self-fertilization can lead to similar difficulties, due to a rapid loss of genetic diversity in heavily inbred lines. All of these suggest that selfing or asexuality causes species to become gradually less fit, and thus, presumably, rarer (Figure 1.2a). More formally, we can imagine a geographical fitness landscape for an outcrossing species, with the population's intrinsic growth rate varying across space. If selfing or asexuality causes an overall decrease in performance, it should result in pushing this fitness surface down, causing decreases both in the distribution of the species and in its local abundance. This could cause self-compatible and asexual species to be rarer than average, or, to put it the other way around, **disproportionately high levels of self-compatibility or asexuality among rare taxa.**

(b) Speciation and self-compatibility

Another (very different) way for a species to become rare is by speciation. Imagine two otherwise similar annual plant species, one self-compatible and one self-incompatible. Surrounding the range of each species is a series of small patches of potential habitat, cut off by swathes of uninhabitable terrain. Very occasionally, a propagule from one species or the other reaches one of these patches. If a seed of the self-incompatible plant arrives, it is likely to die childless, as it is exceedingly unlikely that a potential mate will colonize its patch during its short lifetime. A self-compatible colonist, on the other hand, will be less dependent on finding a conspecific fellow-traveller (Baker, 1955). Being able to set viable seeds by itself, it is much more likely to be able to establish a new local, isolated population. That population would have many of the prerequisites for speciation in place: it is isolated genetically, has passed through a narrow genetic bottleneck, and it may face (presumably) somewhat different selective pressures than the parent population. The result would be a bias in speciation patterns, with self-compatible lines more likely to spawn

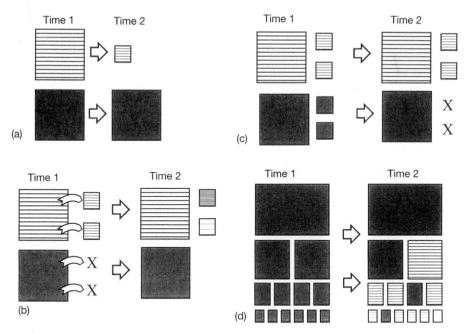

Figure 1.2 Several mechanisms that could each result in rare–common differences in breeding systems. Each box represents a species; the size of a box represents the species' abundance, and its shading represents the species' breeding system (bars = asexual or self-compatible; stippled = self-incompatible or obligately sexual). (a) Selfing as an evolutionary dead end: asexual species grow increasingly unfit, and increasingly rare with time. (b) Selective speciation: isolated propagules from asexual species are more likely to become established and speciate into rare endemics. (c) Extinction proneness: both breeding systems are initially represented by species with a range of abundance values, but rare sexual species are vulnerable to extinction. (d) Transformations: rare sexual species are liable to become asexual over time, due to any of a number of processes (outlined in the text).

numerous (self-compatible) daughter species, many of which would remain as narrowly distributed endemic taxa (Figure 1.2b). The result (again): **disproportionately high levels of self-compatibility or asexuality among rare taxa**.

1.2.2 Exit rules

(a) Extinction proneness

Among self-incompatible species, the reproductive success of individual plants may depend on the local density of conspecifics in their immediate vicinity (Chapter 9; Feinsinger *et al.*, 1991; Kunin, 1993). Both in wind and animal pollinated species, the probability of successfully getting pollen to receptive stigmas and of receiving pollen from one's neighbours drops dramatically if local density is too low. As a result, populations below some critical threshold density may experience a positive feedback loop towards extinction, with ever rarer populations performing ever worse until none remain. On a local scale, this means that self-incompatible species may tend to form high-density patches rather than spreading thinly over the landscape. On a larger spatial scale, areas that might otherwise have been able to support sparse populations of a species may well be left unoccupied, potentially resulting in regional or even global extinction. Thus, even if sexual and asexual species were initially assigned densities at random, over time the rarest obligately sexual species might be weeded out (Figure 1.2c). The result (yet again): **disproportionately high levels of self-compatibility or asexuality among rare taxa.**

1.2.3 Transformations

Finally, we can look at the processes that transform species as a consequence of their abundance. The underlying logic here is the same for a variety of processes: rare species, even if they begin as obligately sexual, gradually develop a proclivity to asexuality and selfing (Figure 1.2d).

(a) Plasticity in breeding systems

Low plant population densities can cause severe declines in pollinator visitation rates and in the effectiveness (in terms of pollen transfer) of those visits received. Even more severe declines in the effectiveness of abiotic (e.g. wind) pollination systems occur as density declines. This may lead to partial or complete reproductive failure in some self-incompatible plants. Many plant species, however, show some degree of plasticity in their allocation of reproductive resources. This may involve mechanisms for delayed self-fertilization, switches from chasmogamous to cleistogamous flowers, or shifts from sexual to vegetative reproduction. If sparsely populated plants are less successful than those with denser populations at achieving cross-pollination, the result would be a *de facto* shift to selfed or asexual reproductive pathways. As a result, a survey of the realized breeding systems of sparsely and densely populated species might reveal **disproportionately high levels of selfing or asexuality among rare taxa**.

(b) Decreased inbreeding depression

The ecological shift in realized breeding systems suggested above would have genetic and evolutionary consequences as well. The shift to greater levels of selfing should result in increasing selective pressure against deleterious recessive alleles, lowering the species' genetic load. This, in turn, should reduce one of the most important selective disadvantages to selfing: inbreeding depression (Lande and Schemske, 1985; Schemske and

Lande, 1985). Thus the relative evolutionary advantages and disadvantages of evolving greater self-compatibility would be affected by the very act of beginning to self-pollinate more. The ecological shift to greater selfing should result in an evolutionary shift to greater self-compatibility. The net effect: **disproportionately high levels of self-compatibility or asexuality among rare taxa**.

(c) Genetic drift

There are more direct genetic consequences of rarity than those discussed above. Perhaps the most striking consequence of small population sizes is an increasing propensity to genetic drift – the loss of genetic diversity from the population by chance. Drift is considered one of the chief reasons for the genetic impoverishment of rare species' populations (Chapter 10), but it may have direct consequences for the evolution of plant breeding systems as well. Many self-incompatible plants rely on a variety of S-alleles to distinguish selfed from outcrossed pollen. Genetic drift in small populations could lead to the loss of many of these alleles, and thus compromise the workability of the system. An increasingly large percentage of pollen in the population would come to share the same S-alleles, making it increasingly difficult to find any sexual partner that would pass muster genetically. Thus significant selective pressures for self-compatibility might arise, not because of any intrinsic advantage to selfing *per se*, but rather because of the increasing impracticality of the mechanism used to prevent it. The result, in any case, is the same: **disproportionately high levels of self-compatibility or asexuality among rare taxa**.

(d) Allee effects and selection

Self-compatibility, however, may have direct evolutionary advantages for rare species to complement the indirect ones cited above. The possibility of reproductive difficulties in low-density plant populations (Allee effects) creates the potential for significant selective pressures to circumvent them. If sparsely populated plants have difficulty attracting effective pollinators, then those individuals least dependent on their visits may enjoy the greatest reproductive success. If there is heritable variation between individuals in the strength of their self-incompatibility systems (and there often is), then those individuals with the greatest capacity for selfing should be at a selective advantage in sparse populations (Chapter 11). Over evolutionary time, the result should be a gradual shift in reproductive systems to fit the local selective regime, leading to **disproportionately high levels of self-compatibility or asexuality among rare taxa**.

In case you have failed to notice the refrain at the end of each of the above paragraphs, let us further belabour the point one last time. Despite the wide array of processes considered, the result in each case is the same: **rare species are disproportionately likely to be self-compatible or asexual**. The point of all this is not so much to demonstrate something about plant breeding systems (although it is the case that first drew my attention to the issue); a similar list of scenarios could be drawn up for a number of other cases (e.g. insect dispersal abilities or bird life-history strategies). Rather, it is trying to illustrate the difficulties involved in moving from an observed pattern to an analysis of the processes that could produce it.

1.3 THE ORGANIZATION OF THE BOOK

This book is designed to explore some of these arguments, and to test their relevance towards explaining rare–common differences, not just in breeding systems but in a variety of species traits. The organization of the book roughly parallels the organization of this introduction: it begins in Part One with the subject of rarity in general and of rare–common differences in particular; Part Two then examines potential mechanisms for generating such patterns (entry, exit and transformative processes in turn) and the book concludes in Part Three with ideas about what is to come.

More specifically, Part One begins with a review of the literature on differences between rare and common species (Chapter 2) and an analysis of some of the issues plaguing the study (and definition) of rarity (Chapters 3 and 4). Part Two commences with a cluster of chapters examining the potential for community assembly rules (Chapter 5) and selective speciation (Chapter 6) to bias the set of species becoming rare – two forms of entry rules. The next two chapters examine the effects of extinction in editing the set of rare species (that is, exit rules), from paleontological (Chapter 7) and recent ecological (Chapter 8) perspectives. The possible consequences of rarity in transforming species characteristics is the focus of the subsequent section: Chapter 9 examines the ecological consequences of rarity, Chapter 10 focuses on genetic effects, and Chapters 11 and 12 consider possible evolutionary responses to rarity, from empirical and theoretical perspectives, respectively. Part Three, the final section of the book, attempts to bring these disparate perspectives together and point to fruitful new directions for research. Chapter 13 considers new comparative methods which may make it possible to distinguish the causes of rarity from its consequences. The book is then concluded with a summary of some of the salient points raised and of important open questions for future work (Chapter 14).

REFERENCES

Baker, H.G. (1955) Self-compatibility and establishment after 'long-distance' dispersal. *Evolution*, **9**, 347–348.

Feinsinger, P., Tiebout, H.M. III and Young, B.E. (1991) Do tropical bird-pollinated plants exhibit density-dependent interactions? Field experiments. *Ecology*, **72**, 1953–1963.

Gaston, K.J. (1994) *Rarity*, Chapman & Hall, London.

Hamilton, C.W. (1990) Variations on a distylous theme in Mesoamerican *Psychotria* subgenus *Psychotria* (Rubiaceae). *Mem. N.Y. Bototanical Garden*, **55**, 62–75.

Jablonski, D. (1987) Heritability at the species level: Analysis of geographic ranges of Cretaceous mollusks. *Science*, **238**, 360–363.

Karron, J.D. (1987) A comparison of levels of genetic polymorphism and self-compatibility in geographically restricted and widespread plant congeners. *Evolutionary Ecology*, **1**, 47–58.

Kunin, W.E. (1991) Few and far between: plant population density and its effects on insect–plant interactions. PhD thesis, University of Washington.

Kunin, W.E. (1993) Sex and the single mustard: population density and pollinator behavior effects on seed-set. *Ecology*, **74**, 2145–2160.

Kunin, W.E. and Gaston, K.J. (1993) The biology of rarity: patterns, causes and consequences. *Trends in Ecology and Evolution*, **8**, 298–301.

Lande, R. and Schemske, D.W. (1985) The evolution of self-fertilization and inbreeding depression in plants. I. Genetic models. *Evolution*, **39**, 24–40.

Longton, R.E. (1992) Reproduction and rarity in British mosses. *Biological Conservation*, **59**, 89–98.

Müller, H.J. (1964) The relation of recombination to mutational advance. *Mutation Research*, **1**, 2–19.

Schemske, D.W. and Lande, R. (1985) The evolution of self-fertilization and inbreeding depression in plants. II. Empirical observations. *Evolution*, **39**, 41–52.

2 Rare–common differences: an overview

Kevin J. Gaston and William E. Kunin

2.1 INTRODUCTION

Comparative studies of the biological traits of taxonomically related rare and common species are scarce. The past couple of decades, however, have seen a growing number of attempts to rectify this situation. In the main (though not exclusively), these studies have concerned small numbers of species, in one or a few genera, and have examined at most a few features of their biologies. Nonetheless, some potentially important regularities have begun to emerge. In this chapter we will consider several of these patterns and review the evidence for them. Discussion of possible mechanisms and related theory is left to later chapters in the book (many of the studies cited here will thus make later reappearances).

In reviewing the empirical work on rare–common differences it is important to recognize its limitations. The work has usually been non-experimental and has been able to conclude little about the causes of observed differences. Moreover, much of it has been severely complicated both by multiple statistical interactions between different traits and by statistical artefacts (Gaston, 1994). An additional complication stems from the fact that rarity has been defined in a variety of different ways and at a range of spatial scales, and, partly as a result, the forms which analyses take are at best heterogeneous. In consequence, the results of any two studies are seldom directly comparable in any but the broadest qualitative sense (Chapter 3; Kunin and Gaston, 1993). Whilst in the longer term it would seem a potentially valuable approach, meta-analysis and related techniques (e.g. Tonhasca and Byrne, 1994) cannot at present be applied to quantify rigorously the weight of evidence for any given pattern. The potential complications of the phylogenetic relatedness of species (Chapter 13; Harvey and Pagel, 1991) have seldom been addressed, raising the possibility that some documented patterns are in fact artefacts. Finally, the published literature is not an unbiased sample of analyses which have been

The Biology of Rarity. Edited by William E. Kunin and Kevin J. Gaston. Published in 1997 by Chapman & Hall, London. ISBN 0 412 63380 9.

performed: those studies that find significant differences in the biological traits of rare and common species probably have a greater likelihood both of being written up and of being published.

With these caveats in mind, we will here address rare–common differences in eight groups of traits: breeding systems, reproductive investment, dispersal ability, homozygosity, competitive ability, resource usage, trophic status and body size. Although we will not dwell on them, a number of additional rare–common differences which cannot strictly be viewed as species traits have also been explored in the literature. These include, for example, consumer loads, numbers of competitors and numbers of coexisting species (e.g. Lawton and Price, 1979; Kennedy and Southwood, 1984; Strong et al., 1984; Leather, 1986; Brown and Kurzius, 1987; Claridge and Evans, 1990; Gregory, 1990; Gregory et al., 1991; Aho and Bush, 1993; Straw, 1994; Straw and Ludlow, 1994; Durrer and Schmid-Hempel, 1995). The terms 'rare', 'common', 'abundance' and 'range size' are used in a broad sense throughout the chapter, and are not constrained to any particular spatial scale. Nonetheless, we will frequently specify the form of rarity considered by those studies to which we refer.

2.2 RARE–COMMON DIFFERENCES IN PARTICULAR TRAITS

2.2.1 Breeding system

There have been a number of studies comparing the reproductive characteristics of rare and common species, especially of plants. Overall, they suggest rare species rely less on outcrossing and sexual reproduction in general than do their commoner counterparts. Longton (1992) found that, among the nearly 700 species in the British moss flora, regionally and globally rare species were disproportionately likely to be non-fruiting and thus dependent on vegetative reproduction. Among those species that did produce sporophytes, rare species were almost always monoecious, and so at least potentially capable of self-fertilization, whereas a large percentage of common species were dioecious and thus obligate outcrossers. Similarly, for mosses in a boreal Canadian national park, Hedderson (1992) found that species presumed to reproduce by asexual fragmentation were significantly over-represented among the rare group (those at the edge of their geographical ranges).

Similar patterns show up among vascular plants. Hamilton (1990) surveyed herbarium specimens of some 66 species of Central American shrubs of the genus *Psychotria*. The genus is composed largely of heterostylous taxa, where individuals are presumably not only self-incompatible, but mutually incompatible with other plants sharing their floral morphology. Hamilton found that narrowly distributed species were far more likely than their widespread congeners to have monomorphic (or even homostylous) flowers, strongly suggesting a breakdown of their incompatibility system. Orians (Chapter 11) has carried out further work on this genus, confirming the link between the loss of heterostyly and the breakdown of self-incompatibility. Karron (1987a) also compared wide-spread and geographically restricted congeners, compiling data on 11 genera. Even though many of the genera, including Karron's own (1987b) studies of three *Astragalus* spp., showed higher self-compatibility in the more restricted species, others showed the opposite and no overall pattern emerged.

Where outcrossing is required, rare plants may need more focused pollen vectors than their commoner counterparts. Harper (1979), in a comparison of endangered plant lists and their associated floras, found wind pollination (which is inefficient at pollen transfer at low density) to be under-represented among rare species. He also found zygomorphy (bilateral rather than radial symmetry in flowers), which is often associated with specialized pollinator services and efficient pollen transfer, to be over-represented among rare species. Kunin and Shmida (in press) found a rather subtler pattern in their analysis of annual crucifers. They examined the plants' breeding systems and a number of floral traits as functions of the abundance of the species (on three spatial scales), and found sparsely populated species to be self-compatible more often than would be expected by chance. They also found significant interactions between abundance and breeding system in their analyses of many floral characteristics. Petal size, for example, was generally greater in obligate outcrossers than in self-compatible species. Among rare species, however, the patterns were exaggerated; they had unusually small petals if they were self-compatible and unusually large petals if they were self-incompatible.

The preponderance of plant studies among breeding system comparisons is probably due to the fact that plants display a wider variety of breeding systems than do most 'higher' animal groups. In those animal groups that are not obligately sexual, similar tendencies for rare species to be more disproportionately asexual or self-compatible are to be expected. Sonneborn (1957), for example, found rare *Paramecium* varieties dis-proportionately likely to be autogamous. Similarly, nearly all cases of asexuality known among terrestrial vertebrates come from taxa with extremely narrow distributions. More studies of the breeding systems of rare and common animals are needed, especially among groups where asexuality or hermaphroditism are widespread. Among groups with nearly universal sexuality, rare and common species may still differ in effective rates of outcrossing, due to differences in dispersal patterns.

2.2.2 Reproductive investment

Rare–common differences have been reported in other characteristics of the reproductive biology of species in addition to breeding systems, most notably in levels of investment in reproduction (e.g. Meagher *et al.*, 1978; Glazier, 1980; Paine, 1990). Paine (1990) found that species of darter fishes that were rare (in terms of size of geographical range) tended not to put as much energy into reproduction as did widespread species. Likewise, North American *Peromyscus* mice with narrowly restricted geographical ranges had smaller average litter sizes and smaller maximum litter sizes, exhibited less reproductive potential (in the laboratory), and expended less reproductive effort (estimated energy cost of raising a single litter) than their commoner congeners (Glazier, 1980). Again there are counter-examples, including those of Laurance (1991) who found no relation-ship between the abundance of Australian tropical rain forest mammal species and their fecundity (typical number of young produced annually).

Sutherland and Baillie (1993) found a positive relationship between potential fecundity (mean clutch-size × mean number of broods per season) and range size among British bird species, but no significant interaction between potential fecundity and density. Similarly, potential population growth rate – calculated as $[\ln(F/2)]/T$, where F is the egg number per female and T is the mean generation time – and geographical range size have been found to be positively related for noctuid moths (Spitzer and Lepš, 1988, 1992; Inkinen, 1994), though population growth rate was not in general correlated with abundance (Spitzer and Lepš, 1988; Inkinen, 1994). On the other hand, among bactivorous ciliates, rare species apparently have fast population growth for their body size (Taylor, 1978).

The extent to which lower levels of reproductive investment made by rare species per unit time may in some groups translate into lower levels of lifetime reproductive investment is unclear. The lower investment dis-played may reflect a trade-off with longevity. Thus, Glazier (1980) found that although rare species of *Peromyscus* had lower reproductive rates, they also lived longer.

In the extreme, rare species may fail to reproduce in an area at all. In the Sheffield flora a smaller proportion of rare species than common species have readily germinating seeds, or a capacity for vegetative lateral spread, whilst a higher proportion produce no viable seed in the study region (Hodgson, 1986). However, Rabinowitz *et al.* (1989) found that sparse grass species showed lower variation in culm production and seed-set, and were less likely to show reproductive failure than were common grasses in the same habitat. Other examples are also known in which features of reproductive investment are not associated with abundance or spatial distribution; Maly (1991) found little relationship among clutch size, incidence and distribution patterns of Australasian freshwater centropagid copepods.

Based on theoretical arguments which frequently, in part at least, concern differences in reproductive rates, there has been much discussion of possible differences in the temporal dynamics of populations of rare and common species (e.g. Glazier, 1986; Gaston and Lawton, 1988; Hanski, 1990; McArdle and Gaston, 1992; Gaston and McArdle, 1994). Such differences essentially lie outside the scope of this volume and have proved difficult to test rigorously (Gaston and McArdle, 1994).

2.2.3 Dispersal ability

In general, studies have concluded that rare species have poorer dispersal abilities than common (in this case rarity is almost invariably expressed in terms of geographical range size: Hansen, 1980; Reaka, 1980; Juliano, 1983; Kavanaugh, 1985; Söderström, 1987, 1989; Hedderson, 1992; Oakwood et al., 1993; Novotný, 1995). For example, Kavanaugh (1985) found that the mean range size of brachypterous (wingless) species of Nearctic carabid beetles of the genus *Nebria* was only 14% of that of macropterous species. Indeed, among beetles in general, flightless species often have restricted geographical ranges (Crowson, 1981).

Söderström (1987) found that for epixylic mosses in late successional spruce forests, species occurring in small populations at a few of the available localities produce diaspores only occasionally and are interpreted to have poor dispersal ability, even within localities, or establishment difficulties. Similarly, the disproportionate reliance of rare mosses on vegetative reproduction (Hedderson, 1992; Longton, 1992), noted above, suggests poorer dispersal abilities in such taxa.

Such relationships may often be asymmetric, with good dispersers having small, intermediate and large range sizes, but poor dispersers tending only to have small to intermediate-sized ranges. In parallel with the known decline in the mean geographical range sizes of species from high to low latitudes, there is some evidence for a decline in the dispersal abilities of species, at least for plants (Huston, 1994).

Dispersal abilities are difficult to assess, and most comparative studies addressing their association with abundance and range size have relied on rather indirect measures. This may be problematic. For example, Rabinowitz (Rabinowitz, 1978; Rabinowitz and Rapp, 1981) found that rare prairie grass species have lighter dispersal units than common species, which resulted in a greater dispersal distance when measured in the laboratory. In the field, however, systematic differences in dispersal ability between rare and common species were not found, largely because of interspecific differences in the height from which the diaspores were released.

The pattern of smaller seeds among rare taxa has also been noted among plant species in chalk grassland (Mitchley and Grubb, 1986) and North American oaks (Aizen and Patterson, 1990; but see Jensen, 1992; Aizen and Patterson, 1992). In contrast, seed weight in sand dune annuals has been found to be negatively related to the proportional abundance of a species within a community, average abundance in occupied quadrats and the proportion of quadrats occupied (Rees, 1995). Oakwood *et al.* (1993) found a weak tendency in the Australian flora for species with larger diaspores to have narrower geographical ranges. Seed size appears to be positively linked with competitive ability in this system, and there is some evidence for a colonization–competition trade-off (larger-seeded species being poorer colonists but better competitors).

Assessment of rare–common differences in dispersal ability also provides good examples of the complications of multiple interactions. Thus, Peat and Fitter (1994) found that, for British angiosperms, wind-dispersed species (presumably with greater dispersal abilities) have more restricted geographical ranges than do species with unspecialized dispersal mechanisms. However, wind-dispersed species tend to produce greater numbers of smaller seeds, and there is a general positive correlation between seed production and geographical range; the most wide-ranging species are those that produce many seeds and have unspecialized dispersal mechanisms. Other studies show that high dispersal abilities due to wind (Gentry, 1992; Kelly *et al.*, 1994) or vertebrate dispersers (Oakwood *et al.*, 1993) are correlated with disproportionately wide geographical ranges. Overall, the interactions between seed size, seed number, reproductive investment, dispersal mode and dispersal ability may be very complex indeed, making any interpretation of rare–common differences difficult.

For terrestrial taxa, interpretation of any tendency for rare species to have poorer dispersal abilities than common is potentially complicated by interactions of both rarity and dispersal ability with environmental stability. Many studies of rare–common differences in dispersal ability in this realm have been performed in temperate landscapes dominated by agricultural activities. These areas often undergo frequent disturbance, and tend therefore to be colonized by species with good dispersal abilities. This may result in an association between rarity and poor dispersal ability which may not be evident under other circumstances.

Finally, it may not always be straightforward to differentiate between the interactions of dispersal ability and of establishment ability (which may in part be associated with reproductive investment) and body size.

2.2.4 Genetic polymorphism

One of the best documented patterns of differences between rare and common species concerns their genetic structure. There is a sizeable theoretical literature on the implications of population size on inbreeding levels and genetic drift – processes likely to lead to losses of genetic diversity in rare populations. Moreover, if rare and common species differ in breeding systems and dispersal abilities, it is hardly surprising that they differ in genetic properties as well. There are, of course, some documented cases of common species which have very low levels of genetic polymorphism and of rare species with moderate levels (Karron, 1987a; Baur and Schmid, 1996), and cases in which there is little evidence for rare–common differences in genetic variation (e.g. Primack, 1980). However, in general, rare species tend to be genetically impoverished compared with common species (e.g. Karron, 1987a; Moran and Hopper, 1987; Sherman-Broyles *et al.*, 1992). Most studies to date have considered rarity and commonness exclusively in terms of geographical range size; a more complete analysis of the genetic correlates of rarity on a variety of spatial scales is given in Chapter 10.

2.2.5 Competitive ability

The notion that competitive inferiority typifies many rare species has long been held (e.g. Griggs, 1940). This idea has recently come under experimental scrutiny for plants. In agreement with much current theory, the results indicate no consistent pattern. Some studies report a direct relationship between relative abundance and competitive ability (e.g. Grubb, 1982; Mitchley and Grubb, 1986; Miller and Werner, 1987). Other studies have found no such interaction (e.g. Rabinowitz *et al.*, 1984; Taylor and Aarssen, 1990; Duralia and Reader, 1993). There are difficulties in interpreting many of these studies, as abundance is often measured in terms of biomass, a measure that gives large species – which may be competitively dominant on account of their stature – an advantage (Silvertown and Dale, 1991). One might expect rare species to be better adapted to interspecific rather than to intraspecific competition, as widely scattered individuals will tend to be surrounded by their commoner neighbours. For example, in greenhouse experiments Rabinowitz *et al.* (1984) found that species of prairie grasses which were sparse (large geographical ranges, but never abundant) overyielded and were advantaged by interaction with common species in greenhouse experiments. Also working on prairie grasses (three species), Duralia and Reader (1993) reported field experiments which indicated that competitive ability and abundance were not significantly related.

There has been surprisingly little work on the relationship between competitive ability and abundance in animals. Ritchie (in preparation) experimentally measured the competitive abilities of grasshopper species, and found a significant positive relation with relative abundance in most of the 25 fields he analysed. More studies of this kind are badly needed.

The above said, study of the interaction between abundance or range size and competitive ability is severely constrained, particularly in the field, by the potentially seriously confounding effects of succession and disturbance regime, as well as size effects (as noted above). Furthermore, the relationship between competitive ability and abundance may depend on the way in which communities are organized (Chapter 5). If species have distinct preference niches that overlap at their edges, competitive dominants may be able to 'realize' a greater fraction of their fundamental niches, and thus be more abundant. With shared preference niches, however, the competitively dominant species will also tend to be the species with the narrowest tolerances, and thus may be rare. This leads us, naturally enough, to the subject of specialization.

2.2.6 Resource usage

Rare and common species may arguably differ systematically in their patterns of resource usage (defined broadly to embrace sets of environmental conditions, and habitats, as well as nutrition) in two ways (Gaston, 1994). First, rare species may utilize resources which themselves occur at lower abundances or over more restricted areas than do those resources used by common species; rare species may be described as, in this sense, exhibiting habitat peculiarity (Prober and Austin, 1991; Pate and Hopper, 1993). Second, rare species may utilize a narrower range of resources than do common species; they have narrower niche breadths.

Comparative studies of resource usage are complicated, or rendered more difficult to interpret, by sample size and various other effects (Gaston, 1994). For example, rare species may artefactually appear to utilize a narrower range of resources than common species, if analyses are based on the resource usage of fewer individuals of the former than of the latter. Taking the results at face value, however, there is support for rare–common differences in resource usage of both forms (Gaston, 1994). Rare species have been reported to utilize resources which occur at lower abundances or over more restricted areas (e.g. Seagle and McCracken, 1986; Dixon and Kindlmann, 1990; Gilbert, 1991; Dixon, 1994), and to use a narrower range of resources (e.g. Hanski and Koskela, 1978; Owen and Gilbert, 1989; Pomeroy and Ssekabiira, 1990; Hedderson, 1992; Shkedy and Safriel, 1992; Hodgson, 1993; Virkkala, 1993; Taylor et al., 1993; Inkinen, 1994). Rare–common differences in resource usage are frequently complex, with empirical studies providing only limited support for some interactions, with counter-examples existing (Gaston, 1994) and with theory increasingly predicting differences between the relationships between resource usage and abundance as against relationships between resource usage and spatial distribution (Ford, 1990; Kouki and Häyrinen, 1991; Hanski et al., 1993).

Along with several other groups of traits (e.g. homozygosity, reproductive investment), there are strong parallels between hypothesized differences in the resource usage of common and rare species and suggested differences between core and marginal populations (core populations are often larger than marginal ones) of individual species (Hengeveld and Haeck, 1981; Brown, 1984; Hengeveld, 1990). This raises the possibility that similar processes are at work in these different circumstances.

2.2.7 Trophic group

Typically the average abundance of animal species tends to be lower at high levels of a trophic hierarchy (e.g. Elton, 1927; Eisenberg, 1980; Arita *et al.*, 1990; Laurance, 1991; Thiollay, 1994; but see Chapter 5). For example, mammalian primary consumers have higher densities than secondary consumers (Damuth, 1987), and the minimum viable densities of herbivorous mammals are 13 times those of carnivores and insectivores (Silva and Downing, 1994). In general, differences appear most marked between the abundances of carnivores and non-carnivores, with little repeatable distinction between the average abundances of many other trophic groups.

Whilst big fierce animals tend to be rare (Colinvaux, 1980) in terms of their abundances, this is not necessarily so with respect to their geographical range sizes. Among the mammals at least, species at higher trophic levels have large geographical range sizes; those of Carnivora tend to be larger, on average, than those of species in other terrestrial mammalian orders (e.g. Brown, 1981; Rapoport, 1982; Pagel *et al.*, 1991; Letcher and Harvey, 1994).

Amongst host-specialist consumers, geographical range sizes decline at higher levels of a trophic hierarchy, because a consumer species seldom occurs throughout the range of its host (for examples see Strong *et al.*, 1984; Pekkarinen and Teräs, 1993; Dixon, 1994; Whitcomb *et al.*, 1994). The extent to which consumers more broadly (generalists as well as specialists) tends to have smaller geographical ranges than their hosts is not known.

In most cases there is substantial overlap in the magnitudes of the densities and the sizes of the geographical ranges of species belonging to different trophic groups. Interpretation of interactions between abundance or range size and trophic status may be complicated by the frequent association between trophic status and body size (e.g. fierce animals tend to be big; Chapter 5).

2.2.8 Body size

Of all rare–common differences, those in body size arguably have attracted the greatest attention. This interest has been directed principally toward relationships between abundance and body size in animal assemblages, and has yielded two groups of results (for reviews see Cotgreave, 1993; Blackburn and Lawton, 1994). First, a number of studies have found linear relationships on logarithmic axes, with smaller-bodied species having higher densities (Damuth, 1981, 1987; Peters and Wassenberg, 1983; Peters and Raelson, 1984; Arita *et al.*, 1990; Currie, 1993; Currie and Fritz, 1993). These studies have tended to be based on compendia of data drawn from the literature, and may as a result be biased in important ways (Lawton, 1989, 1991). The second group of studies have been based on sampling whole assemblages of taxonomically similar animals, and have generally revealed polygonal relationships between abundance and body size (again on a logarithmic scale), with intermediate-sized species being the most abundant and rare species having a range of body sizes (e.g. Gaston, 1988; Morse *et al.*, 1988; Carrascal and Tellería, 1991; Blackburn *et al.*, 1993; Cotgreave *et al.*, 1993; Blackburn and Lawton, 1994; Cambefort, 1994). Indeed, using data of this kind, positive relationships between abundance and body size are not unknown (e.g. Gaston *et al.*, 1993; Nilsson *et al.*, 1994), perhaps disproportionately within taxa of low rank (e.g. genera and tribes as opposed to families or orders; Nee *et al.*, 1991, but see also Cotgreave and Harvey, 1992, 1994; Blackburn *et al.*, 1994; Cotgreave, 1994).

A number of studies have documented broadly positive relationships between the geographical range size and body size of animal species (e.g. Van Valen, 1973; Reaka, 1980; Brown, 1981; McAllister *et al.*, 1986; Brown and Maurer, 1987, 1989; Arita *et al.*, 1990; Brown and Nicoletto, 1991; Maurer *et al.*, 1991; Ayres and Clutton-Brock, 1992; Cambefort, 1994; Taylor and Gotelli, 1994). Not infrequently, when studies embrace the entire geographical ranges of all or most of the species in an assemblage, these interactions tend to be rather polygonal (though obviously in a different way to abundance–body-size relationships), with species of a spectrum of body sizes having large ranges, but with the lower bound to range sizes increasing with body size (e.g. Brown and Maurer, 1987; Gaston, 1994; Gaston and Blackburn, 1996a). When they are less comprehensive (i.e. their spatial extent embraces only part of the geographical ranges of many of the species in an assemblage), a number of studies report negative interactions between range size and body size (e.g. Sutherland and Baillie, 1993; Cambefort, 1994), and others no significant simple interaction (e.g. Juliano, 1983; Hodgson, 1993; Gaston, 1994; Inkinen, 1994; the effect of spatial scale on range size-body size relationships has been reviewed by Gaston and Blackburn, 1996b). Effects of phylogeny on the direction of relationships between range size and body size have been documented (Glazier, 1980; Taylor and Gotelli, 1994).

Interspecific relationships between abundance and body size and between range size and body size have received little attention in the context of plants, for which the plasticity of growth has been regarded as making body size a less meaningful variable. Nonetheless, some analyses have been performed. Aizen and Patterson (1990) find a positive relationship between the geographical range size and height of North American oaks (*Quercus* species). Harper (1979) argues that larger (woody) plant species are in general under-represented in the rare component of floras, while Oakwood *et al.* (1993) find that plant species of taller growth forms tend to occupy a smaller number of regions of Australia.

2.3 CONCLUSION

With various caveats and probably many exceptions, but across a diverse array of study organisms, it would appear that rare species differ from common in that they:

- have breeding systems biased away from outcrossing and sexual reproduction;
- have lower reproductive investment;
- have poorer dispersal abilities;
- have higher levels of homozygosity;
- use less common resources and/or a narrower range of resources;
- have, for abundances, a greater probability of belonging to groups at higher levels of a trophic hierarchy;
- have, for abundances, under some circumstances, a greater probability of a larger body size and, for geographical range sizes, a greater probability of a smaller body size.

These differences constitute broad generalizations, often with substantial unexplained variance and only moderate explanatory power. They cannot be regarded as identifying a discrete set of rare species characterized by a suite of consistent traits. Nonetheless, these patterns remain intriguing, and of potential importance, both heuristic and applied. They may arise for a number of reasons, which the rest of this book attempts to evaluate.

ACKNOWLEDGEMENTS

K.J.G. is a Royal Society University Research Fellow. We are grateful to Tim Blackburn, Mick Crawley, Mark Rees and Phil Warren for helpful discussion.

REFERENCES

Aho, J.M. and Bush, A.O. (1993) Community richness in parasites of some freshwater fishes from North America, in *Species Diversity in Ecological Communities: historical and geographical perspectives* (eds R.E. Ricklefs and D. Schluter), University of Chicago Press, Chicago, pp. 185–193.

Aizen, M.A. and Patterson III, W.A. (1990) Acorn size and geographical range in the North American oaks (*Quercus* L.). *Journal of Biogeography*, 17, 327–332.

Aizen, M.A. and Patterson III, W.A. (1992) Do big acorns matter? – a reply to R.J. Jensen. *Journal of Biogeography*, 19, 581–582.

Arita, H.T., Robinson, J.G. and Redford, K.H. (1990) Rarity in Neotropical forest mammals and its ecological correlates. *Conservation Biology*, 4, 181–192.

Ayres, J.M. and Clutton-Brock, T.H. (1992) River boundaries and species range size in Amazonian primates. *American Naturalist*, 140, 531–537.

Baur, B. and Schmid, B. (1996) Spatial and temporal patterns of genetic diversity among species, in *Biodiversity: a biology of numbers and difference* (ed. K.J. Gaston), Blackwell Scientific, Oxford, pp. 169–201.

Blackburn, T.M. and Lawton, J.H. (1994) Population abundance and body size in animal assemblages. *Philosophical Transactions of the Royal Society, London, B*, **343**, 33–39.

Blackburn, T.M., Brown, V.K., Doube, B. *et al.* (1993) The relationship between body size and abundance in natural animal assemblages. *Journal of Animal Ecology*, **62**, 519–528.

Blackburn, T.M., Gates, S., Lawton, J.H. and Greenwood, J.J.D. (1994) Relations between body size, abundance and taxonomy of birds wintering in Britain and Ireland. *Philosophical Transactions of the Royal Society, London, B*, **343**, 135–144.

Brown, J.H. (1981) Two decades of homage to Santa Rosalia: toward a general theory of diversity. *American Zoologist*, **21**, 877–888.

Brown, J.H. (1984) On the relationship between abundance and distribution of species. *American Naturalist*, **124**, 255–279.

Brown, J.H. and Kurzius, M.A. (1987) Composition of desert rodent faunas: combinations of coexisting species. *Annales Zoologici Fennici*, **24**, 227–237.

Brown, J. H. and Maurer, B.A. (1987) Evolution of species assemblages: effects of energetic constraints and species dynamics on the diversification of the American avifauna. *American Naturalist*, **130**, 1–17.

Brown, J.H. and Maurer, B.A. (1989) Macroecology: the division of food and space among species on continents. *Science*, **243**, 1145–1150.

Brown, J.H. and Nicoletto, P.F. (1991) Spatial scaling of species composition: body masses of North American land mammals. *American Naturalist*, **138**, 1478–1512.

Cambefort, Y. (1994) Body size, abundance, and geographical distribution of Afrotropical dung beetles (Coleoptera: Scarabaeidae). *Acta Oecologica*, **15**, 165–179.

Carrascal, L.M. and Tellería, J.L. (1991) Bird size and density: a regional approach. *American Naturalist*, **138**, 777–784.

Claridge, M.F. and Evans, H.F. (1990) Species–area relationships: relevance to pest problems of British trees?, in *Population Dynamics of Forest Insects* (eds A.D. Watt, S.R. Leather, M.D. Hunter and N.A.C. Kidd), Intercept, Andover, Hampshire, pp. 59–69.

Colinvaux, P. (1980) *Why Big Fierce Animals Are Rare*, Penguin Books, Harmondsworth, UK.

Cotgreave, P. (1993) The relationship between body size and abundance in animals. *Trends in Ecology and Evolution*, **8**, 244–248.

Cotgreave, P. (1994) The relationship between body size and abundance in a bird community: the effects of phylogeny and competition. *Proceedings of the Royal Society, London, B*, **256**, 147–149.

Cotgreave, P. and Harvey, P.H. (1992) Relationships between body size, abundance and phylogeny in bird communities. *Functional Ecology*, **6**, 248–256.

Cotgreave, P. and Harvey, P.H. (1994) Phylogeny and the relationship between body size and abundance in bird communities. *Functional Ecology*, **8**, 219–228.

Cotgreave, P., Hill, M.J. and Middleton, D.A.J. (1993) The relationship between body size and population size in bromeliad tank faunas. *Biological Journal of the Linnean Society*, **49**, 367–380.

Crowson, R.A. (1981) *The Biology of the Coleoptera*, Academic Press, London.

Currie, D.J. (1993) What shape is the relationship between body size and population density? *Oikos*, **66**, 353–358.

Currie, D.J. and Fritz, J. (1993) Global patterns of animal abundance and species energy use. *Oikos*, **67**, 56–68.

Damuth, J. (1981) Population density and body size in mammals. *Nature*, **290**, 699–700.

Damuth, J. (1987) Interspecific allometry of population density in mammals and other animals: the independence of body mass and population energy-use. *Biological Journal of the Linnean Society*, **31**, 193–246.

Dixon, A.F.G. (1994) Individuals, populations and patterns, in *Individuals, Populations and Patterns in Ecology* (eds S.R. Leather, A.D. Watt, N.J. Mills and K.R.A. Walters), Intercept, Andover, Hampshire, pp. 449–476.

Dixon, A.F.G. and Kindlmann, P. (1990) Role of plant abundance in determining the abundance of herbivorous insects. *Oecologia*, **83**, 281–283.

Duralia, T.E. and Reader, R.J. (1993) Does abundance reflect competitive ability? A field test with three prairie grasses. *Oikos*, **68**, 82–90.

Durrer, S. and Schmid-Hempel, P. (1995) Parasites and the regional distribution of bumblebee species. *Ecography*, **18**, 114–122.

Eisenberg, J.F. (1980) The density and biomass of tropical mammals, in *Conservation Biology: an evolutionary–ecological perspective* (eds M.E. Soulé and B.A. Wilcox), Sinauer Associates, Sunderland, Massachusetts, pp. 35–55.

Elton, C. (1927) *Animal Ecology*, Sidgwick and Jackson, London.

Ford, H.A. (1990) Relationships between distribution, abundance and foraging specialization in Australian landbirds. *Ornis Scandinavica*, **21**, 133–138.

Gaston, K.J. (1988) Patterns in the local and regional dynamics of moth populations. *Oikos*, **53**, 49–57.

Gaston, K.J. (1994) *Rarity*, Chapman & Hall, London.

Gaston, K.J. and Blackburn, T.M. (1996a) Some conservation implications of geographic range size–body size relationships. *Conservation Biology*, **10**, 638–646.

Gaston, K.J. and Blackburn, T.M. (1996b) Range size–body size relationships: evidence of scale dependence. *Oikos*, **75**, 479–485.

Gaston, K.J. and Lawton, J.H. (1988) Patterns in body size, population dynamics and regional distribution of bracken herbivores. *American Naturalist*, **132**, 662–680.

Gaston, K.J. and McArdle, B.H. (1994) The temporal variability of animal abundances: measures, methods and patterns. *Philosophical Transactions of the Royal Society, London, B*, **345**, 335–358.

Gaston, K.J., Blackburn, T.M., Hammond, P.M. and Stork, N.E. (1993) Relationships between abundance and body size – where do tourists fit? *Ecological Entomology*, **18**, 310–314.

Gentry, A.H. (1992) Tropical forest biodiversity: distributional patterns and their conservation significance. *Oikos*, **63**, 19–28.

Gilbert, L.E. (1991) Biodiversity of a Central American Heliconius community: pattern, process and problems, in *Plant–animal Interactions: evolutionary ecology in tropical and temperate regions* (eds P.W. Price, T.M. Lewinsohn, G.W. Fernandes and W.W. Benson), Wiley, New York, pp. 403–427.

Glazier, D.S. (1980) Ecological shifts and the evolution of geographically restricted species of North American *Peromyscus* (mice). *Journal of Biogeography*, **7**, 63–83.

Glazier, D.S. (1986) Temporal variability of abundance and the distribution of species. *Oikos*, **47**, 309–314.

Gregory, R.D. (1990) Parasites and host geographic range as illustrated by waterfowl. *Functional Ecology*, **4**, 645–654.

Gregory, R.D., Keymer, A.E. and Harvey, P.H. (1991) Life history, ecology and parasite community structure in Soviet birds. *Biological Journal of the Linnean Society*, **43**, 249–262.

Griggs, R.F. (1940) The ecology of rare plants. *Bulletin of the Torrey Botanical Club*, **67**, 575–594.

Grubb, P.J. (1982) Control of relative abundance in roadside *Arrhenatheretum*: results of a long-term experiment. *Journal of Ecology*, **70**, 845–861.

Hamilton, C.W. (1990) Variations on a distylous theme in Mesoamerican *Psychotria* subgenus *Psychotria* (Rubiaceae). *Memoirs of the New York Botanical Garden*, **55**, 62–75.

Hansen, T.A. (1980) Influence of larval dispersal and geographic distribution on species longevity in neogastropods. *Paleobiology*, **6**, 193–207.

Hanski, I. (1990) Density dependence, regulation and variability in animal populations. *Philosophical Transactions of the Royal Society, London, B,* **330**, 141–150.

Hanski, I. and Koskela, H. (1978) Stability, abundance, and niche width in the beetle community inhabiting cow dung. *Oikos*, **31**, 290–298.

Hanski, I., Kouki, J. and Halkka, A. (1993) Three explanations of the positive relationship between distribution and abundance of species, in *Species Diversity in Ecological Communities: Historical and Geographical Perspectives* (eds R. Ricklefs and D. Schluter), University of Chicago Press, Chicago, pp. 108–116.

Harper, K.T. (1979) Some reproductive and life history characteristics of rare plants and implications of management. *Great Basin Naturalist Memoirs*, **3**, 129–137.

Harvey, P.H. and Pagel, M.D. (1991) *The Comparative Method in Evolutionary Biology*, Oxford University Press, Oxford.

Hedderson, T.A. (1992) Rarity at range limits; dispersal capacity and habitat relationships of extraneous moss species in a boreal Canadian National Park. *Biological Conservation*, **59**, 113–120.

Hengeveld, R. (1990) *Dynamic Biogeography*, Cambridge University Press, Cambridge.

Hengeveld, R. and Haeck, J. (1981) The distribution of abundance. II. Models and implications. *Proceedings of the Koninklijke Nederlandse Akademie van Wetenschappen*, **C84**, 257–284.

Hodgson, J.G. (1986) Commonness and rarity in plants with special reference to the Sheffield flora. Part II: The relative importance of climate, soils and land use. *Biological Conservation*, **36**, 253–274.

Hodgson, J.G. (1993) Commonness and rarity in British butterflies. *Journal of Applied Ecology*, **30**, 407–427.

Huston, M.A. (1994) *Biological Diversity: the coexistence of species on changing landscapes*, Cambridge University Press, Cambridge.

Inkinen, P. (1994) Distribution and abundance in British noctuid moths revisited. *Annales Zoologici Fennici*, **31**, 235–243.

Jensen, R.J. (1992) Acorn size redux. *Journal of Biogeography*, **19**, 573–579.

Juliano, S.A. (1983) Body size, dispersal ability, and range size in North American species of *Brachinus* (Coleoptera: Carabidae). *Coleopterists Bulletin*, **37**, 232–238.

Karron, J.D. (1987a) A comparison of levels of genetic polymorphism and self-compatability in geographically restricted and widespread plant congeners. *Evolutionary Ecology*, **1**, 47–58.

Karron, J.D. (1987b) The pollination ecology of co-occurring geographically restricted and widespread species of *Astragalus* (Fabaceae). *Biological Conservation*, **39**, 179–193.

Kavanaugh, D.H. (1985) On wing atrophy in carabid beetles (Coleoptera: Carabidae), with special reference to Nearctic *Nebria*, in *Taxonomy, Phylogeny and Zoogeography of Beetles and Ants* (ed. G.E. Ball), Junk, Dordrecht. pp. 408–431.

Kelly, D.L., Tanner, E.V.J., Nic Lughadha, E.M. and Kapos, V. (1994) Floristics and biogeography of a rain forest in the Venezuelan Andes. *Journal of Biogeography*, **21**, 421–440.

Kennedy, C.E.J. and Southwood, T.R.E. (1984) The number of species of insects associated with British trees: a re-analysis. *Journal of Animal Ecology*, **53**, 455–478.

Kouki, J. and Häyrinen, U. (1991) On the relationship between distribution and abundance in birds breeding on Finnish mires: the effect of habitat specialization. *Ornis Fennica*, **68**, 170–177.

Kunin, W.E. and Gaston, K.J. (1993) The biology of rarity: patterns, causes and consequences. *Trends in Ecology and Evolution*, **8**, 298–301.

Kunin, W.E. and Shmida, A. (in press) Plant reproductive traits as a function of local, regional and global abundance. *Conservation Biology*.

Laurance, W.F. (1991) Ecological correlates of extinction proneness in Australian tropical rain forest mammals. *Conservation Biology*, **5**, 79–89.

Lawton, J.H. (1989) What is the relationship between population density and body size in animals? *Oikos*, **55**, 429–434.

Lawton, J.H. (1991) Species richness and population dynamics of animal assemblages. Patterns in body-size: abundance space. *Philosophical Transactions of the Royal Society, London, B*, **330**, 283–291.

Lawton, J.H. and Price, P.W. (1979) Species richness of parasites on hosts: agromyzid flies on the British Umbelliferae. *Journal of Animal Ecology*, **48**, 619–637.

Leather, S.R. (1986) Insect species richness of the British Rosaceae: the importance of host range, plant architecture, age of establishment, taxonomic isolation and species area relationships. *Journal of Animal Ecology*, **55**, 841–860.

Letcher, A.J. and Harvey, P.H. (1994) Variation in geographical range size among mammals of the Palearctic. *American Naturalist*, **144**, 30–42.

Longton, R.E. (1992) Reproduction and rarity in British mosses. *Biological Conservation*, **59**, 89–98.

Maly, E.J. (1991) Dispersal ability and its relation to incidence and geographic distribution of Australian centropagid copepods. *Verhandlungen der Internationalen Vereinigung für Limnologie*, **24**, 2828–2832.

Maurer, B.A., Ford, H.A. and Rapoport, E.H. (1991) Extinction rate, body size, and avifaunal diversity. *Acta XX Congressus Internationalis Ornithologici*, 826–834.

McAllister, D.E., Platania, S.P., Schueler, F.W. *et al.* (1986) Ichthyofaunal patterns on a geographic grid, in *Zoogeography of freshwater fishes of North America* (eds C.H. Hocutt and E.D. Wiley),. Wiley, New York, pp. 17–51.

McArdle, B.H. and Gaston, K.J. (1992) Comparing population variabilities. *Oikos*, **64**, 610–612.

Meagher, T.R., Antonovics, J. and Primack, R. (1978) Experimental ecological genetics in *Plantago*, III. Genetic variation and demography in relation to survival of *Plantago cordata*, a rare species. *Biological Conservation*, **14**, 243–257.

Miller, T.E. and Werner, P.A. (1987) Competitive effects and responses between plant species in a first-year old field community. *Ecology*, **68**, 1201–1210.

Mitchley, J. and Grubb, P.J. (1986) Control of relative abundance of perennials in chalk grassland in southern England. I. Constancy of rank order and results of pot- and field-experiments on the role of interference. *Journal of Ecology*, **74**, 1139–1166.

Moran, G.F. and Hopper, S.D. (1987) Conservation of the genetic resources of rare and widespread eucalypts in remnant vegetation, in *Nature Conservation: the role of remnants of native vegetation* (eds D.A. Saunders, G.W. Arnold, A.A. Burbidge and A.J.M. Hopkins), Surrey Beatty, Sydney, pp. 151–162.

Morse, D.R., Stork, N.E. and Lawton, J.H. (1988) Species number, species abundance and body length relationships of arboreal beetles in Bornean lowland rain forest trees. *Ecological Entomology*, **13**, 25–37.

Nee, S., Read, A.F., Greenwood, J.J.D. and Harvey, P.H. (1991) The relationship between abundance and body size in British birds. *Nature*, **351**, 312–313.

Nilsson, A.N., Elmberg, J. and Sjöberg, K. (1994) Abundance and species richness patterns of predaceous diving beetles (Coleoptera, Dytiscidae) in Swedish lakes. *Journal of Biogeography*, **21**, 197–206.

Novotný, V. (1995) Relationships between the life histories of leafhoppers (Auchenorrhyncha–Hemiptera) and their host plants (Juncaceae, Cyperaceae, Poaceae). *Oikos*, **73**, 33–42.

Oakwood, M., Jurado, E., Leishman, M. and Westoby, M. (1993) Geographic ranges of plant species in relation to dispersal morphology, growth form and diaspore weight. *Journal of Biogeography*, **20**, 563–572.

Owen, J. and Gilbert, F.S. (1989) On the abundance of hoverflies (Syrphidae). *Oikos*, **55**, 183–193.

Pagel, M.P., May, R.M. and Collie, A.R. (1991) Ecological aspects of the geographic distribution and diversity of mammal species. *American Naturalist*, **137**, 791–815.

Paine, M.D. (1990) Life history tactics of darters (Percidae: Etheostomatiini) and their relationship with body size, reproductive behaviour, latitude and rarity. *Journal of Fish Biology*, **37**, 473–488.

Pate, J.S. and Hopper, S.D. (1993) Rare and common plants in ecosystems, with special reference to the south-west Australian flora, in *Biodiversity and Ecosystem Function* (eds E-D. Schulze and H.A. Mooney), Springer-Verlag, Berlin, pp. 293–325.

Peat, H.J. and Fitter, A.H. (1994) Comparative analyses of ecological character-istics of British angiosperms. *Biological Reviews*, **69**, 95–115.

Pekkarinen, A. and Teräs, I. (1993) Zoogeography of *Bombus* and *Psithyrus* in northwestern Europe (Hymenoptera, Apidae). *Annales Zoologici Fennici*, **30**, 187–208.

Peters, R.H. and Raelson, J.V. (1984) Relations between individual size and mammalian population density. *American Naturalist*, **124**, 498–517.

Peters, R.H. and Wassenberg, K. (1983) The effect of body size on animal abundance. *Oecologia*, **60**, 89–96.

Pomeroy, D. and Ssekabiira, D. (1990) An analysis of the distributions of terrestrial birds in Africa. *African Journal of Ecology*, **28**, 1–13.

Primack, R.B. (1980) Phenotypic variation of rare and widespread species of *Plantago*. *Rhodora*, **82**, 87–96.

Prober, S.M. and Austin, M.P. (1991) Habitat peculiarity as a cause of rarity in *Eucalyptus paliformis*. *Australian Journal of Ecology*, **16**, 189–205.

Rabinowitz, D. (1978) Abundance and diaspore weight in rare and common prairie grasses. *Oecologia*, **37**, 213–219.

Rabinowitz, D. and Rapp, J.K. (1981) Dispersal abilities of seven sparse and common grasses from a Missouri prairie. *American Journal of Botany*, **68**, 616–624.

Rabinowitz, D., Rapp, J.K., Cairns, S. and Mayer, M. (1989) The persistence of rare prairie grasses in Missouri: environmental variation buffered by repro-ductive output of sparse species. *American Naturalist*, **134**, 525–544.

Rabinowitz, D., Rapp, J.K. and Dixon, P.M. (1984) Competitive abilities of sparse grass species: means of persistence or cause of abundance. *Ecology*, **65**, 1144–1154.

Rapoport, E.H. (1982) *Areography: geographical strategies of species*, Pergamon, Oxford.

Reaka, M.L. (1980) Geographic range, life history patterns, and body size in a guild of coral-dwelling mantis shrimps. *Evolution*, **34**, 1019–1030.

Rees, M. (1995) Community structure in sand dune annuals: is seed weight a key quantity? *Journal of Ecology*, **83**, 857–864.

Seagle, S.W. and McCracken, G.F. (1986) Species abundance, niche position, and niche breadth for five terrestrial animal assemblages. *Ecology*, **67**, 816–818.

Sherman-Broyles, S.L., Gibson, J.P., Hamrick, J.L. *et al.* (1992) Comparisons of allozyme diversity among rare and widespread *Rhus* species. *Systematic Botany*, **17**, 551–559.

Shkedy, Y. and Safriel, U.N. (1992) Niche breadth of two lark species in the desert and the size of their geographical ranges. *Ornis Scandinavica*, **23**, 89–95.

Silva, M. and Downing, J.A. (1994) Allometric scaling of minimal mammal densities. *Conservation Biology*, **8**, 732–743.

Silvertown, J. and Dale, P. (1991) Competitive hierarchies and the structure of herbaceous plant communities. *Oikos*, **61**, 441–444.

Söderström, L. (1987) Dispersal as a limiting factor for distribution among epixylic bryophytes. *Symposia Biologica Hungarica*, **35**, 475–484.

Söderström, L. (1989) Regional distributon patterns of bryophyte species on spruce logs in northern Sweden. *The Bryologist*, **92**, 349–355.

Sonneborn, T.M. (1957) Breeding systems, reproductive methods, and species problems in *Protozoa*, in *The Species Problem* (ed. E. Mayr). American Association for the Advancement of Science, Washington, pp. 155–324.

Spitzer, K. and Lepš, J. (1988) Determinants of temporal variation in moth abundance. *Oikos*, **53**, 31–36.

Spitzer, K. and Lepš, J. (1992) Bionomic strategies in Lepidoptera, risk of extinction and nature conservation projects. *Nota Lepididopterum*, **Suppl. 4**, 81–85.

Straw, N.A. (1994) Species–area relationships and population dynamics: two sides of the same coin, in *Individuals, Populations and Patterns in Ecology* (eds S.R. Leather, A.D. Watt, N.J. Mills and K.R.A. Walters), Intercept, Andover, Hampshire, pp. 275–286.

Straw, N.A. and Ludlow, A.R. (1994) Small-scale dynamics and insect diversity on plants. *Oikos*, **71**, 188–192.

Strong, D.R., Lawton, J.H. and Southwood, T.R.E. (1984) *Insects on Plants: community patterns and mechanisms*, Blackwell Scientific, Oxford.

Sutherland, W.J. and Baillie, S.R. (1993) Patterns in the distribution, abundance and variation of bird populations. *Ibis*, **135**, 209–210.

Taylor, C.M. and Gotelli, N.J. (1994) The macroecology of *Cyprinella*: correlates of phylogeny, body size, and geographical range. *American Naturalist*, **144**, 549–569.

Taylor, C.M., Winston, M.R. and Matthews, W.J. (1993) Fish species – environment and abundance relationships in a Great Plains river system. *Ecography*, **16**, 16–23.

Taylor, D.R. and Aarssen, L.W. (1990) Complex competitive relationships among genotypes of three perennial grasses: implications for species coexistence. *American Naturalist*, **136**, 305–327.

Taylor, W.D. (1978) Maximum growth rate, size and commonness in a community of bactivorous ciliates. *Oecologia*, **36**, 263–272.

Thiollay, J.-M. (1994) Structure, density and rarity in an Amazonian rainforest bird community. *Journal of Tropical Ecology*, **10**, 449–481.

Tonhasca Jr, A. and Byrne, D.N. (1994) The effects of crop diversification on herbivorous insects: a meta-analysis approach. *Ecological Entomology*, **19**, 239–244.

Van Valen, L. (1973) Body size and numbers of plants and animals. *Evolution*, **27**, 27–35.

Virkkala, R. (1993) Ranges of northern forest passerines: a fractal analysis. *Oikos*, **67**, 218–226.

Whitcomb, R.F., Hicks, A.L., Blocker, H.D. and Lynn, D.E. (1994) Biogeography of leafhopper specialists of the shortgrass prairie: evidence for the roles of phenology and phylogeny in determination of biological diversity. *American Entomologist*, (Spring), 19–35.

3 What is rarity?

Kevin J. Gaston

3.1 INTRODUCTION

The rare hold a curious fascination. A bizarre variety of objects considered to be rare are avidly sought and collected, studied and catalogued, bought and sold. The extent to which this general enthusiasm for rarity is manifest within the field of population and community biology is perhaps debatable. The practical issues of conservation aside, studies tend predominantly to be concerned with species which would not widely be considered as rare (Kunin and Gaston, 1993; Gaston, 1994a). Nonetheless, concepts of rarity are applied extensively in the relevant literatures, and indeed might be argued to pervade them. They are used in contexts as disparate as sampling, community structure and foraging theory. What, however, do we mean by rarity, in the context of population and community biology? At the outset of an exploration of the differences between rare and common organisms, we would do well to explore the definition of the term. In so doing a framework can be generated which provides some common foundation for interpreting the results of different studies and thence for comparing them.

3.2 VARIABLES

A variety of definitions of rarity and viewpoints on what constitutes rarity exist in the biological literature (Mayr, 1963; Drury, 1974, 1980; Harper, 1981; Margules and Usher, 1981; Rabinowitz, 1981; Reveal, 1981; Main, 1982; Cody, 1986; Fiedler, 1986; Rabinowitz *et al.*, 1986; Soulé, 1986; Usher, 1986a, 1986b; Heywood, 1988; Ferrar, 1989; Hanski, 1991; Fiedler and Ahouse, 1992; McCoy and Mushinsky, 1992; Reed, 1992; Batianoff and Burgess, 1993). Probably all regard rare species as in some sense delimited on the basis of one, two or at most a few variables. What those variables are differs greatly; they have included, for example:

- abundance
- range size

The Biology of Rarity. Edited by William E. Kunin and Kevin J. Gaston.
Published in 1997 by Chapman & Hall, London. ISBN 0 412 63380 9.

- habitat specificity (= habitat occupancy)
- temporal persistence (e.g. taxon age)
- threat (probability of, or time to, extinction)
- gene flow
- genetic diversity
- endemism
- taxonomic distinctness.

Most definitions can be placed in a three-dimensional space, of which the axes are the relative weighting of 'biological measures' (e.g. history, taxonomic isolation, abundance), 'threat measures' (e.g. risk of extinction, estimated time to extinction) and 'value measures' (e.g. how special species are). In the context of population and community biology, it would seem reasonable to argue that we are primarily interested in definitions with low components explicitly associated with value and with threat.

Ideas that rare species have special value which in some way transcends their biological characteristics have no role in generating an explicit ecological definition of rarity. Nonetheless, we must recognize that many species which are almost universally conceived to be rare are popularly considered to be of particular value. The giant panda *Ailuropoda melanoleuca* is perhaps the best example.

Likewise, it is important to divorce considerations of threat from definitions of rarity. Failure to do this has been a recurrent problem in conservation biology. In particular, many schemes to categorize species on the basis of the degree of threat to them have used the term 'rare' as one such category, and have led to the interchangeable application of the term with reference to abundance or range size and likelihood of extinction (Munton, 1987; Burgman *et al.*, 1993).

Concentrating on 'biological' measures, broad consensus favours a definition of rarity based on abundance and/or range size, with species of low abundance or small range size being regarded as rare (herein **range size** is treated as a measure of the area of the spatial distribution of a species, with **geographical range size** being that measure for the full (global) breadth of the occurrence of the species). Virtually all definitions of rarity explicitly mention at least one parameter based on abundance or range size.

A third biological parameter common to several definitions of rarity is habitat breadth – the number or variety of habitats in which a species occurs (e.g. Rabinowitz, 1981; Rabinowitz *et al.*, 1986). However, its inclusion tends to be justified on one of two grounds, neither of which seem particularly strong. First, there may be a prejudgement of the causes of rarity. It is widely held that species of low abundances and small range sizes tend to have narrow habitat ranges. Several other variables which have been used to define rarity suffer from a similar problem of prejudging causes; for example, those with a historical component. Second, the inclusion of habitat breadth as a parameter may be tied to issues of threat, with beliefs that species occupying fewer habitats are at greater risk of extinction.

Confusion over distinctions between rarity and endemism can also unduly complicate definitions of the former. Species are endemic to an area if they occur within it and nowhere else. The narrow or local endemic may thus tend to fit the colloquial notion of rarity. However, the term endemism, in its classical biogeographical usage, does not necessarily imply rarity or even small range size (Kruckeberg and Rabinowitz, 1985).

Following Reveal (1981), a general definition of rarity based solely on abundance and range size might read: 'Rarity is merely the current status of an extant organism which, by any combination of biological or physical factors, is restricted either in numbers or area to a level that is demonstrably less than the majority of other organisms of comparable taxonomic entities.'

3.3 AN OPERATIONAL DEFINITION?

The above definition of rarity is far from being operational. There are at least two broad sets of considerations to be addressed.

3.3.1 Abundance and range size

The first problem to be faced is that abundance and range size themselves beg definition. Both can be measured in a variety of ways. This issue has received considerable attention in the context of abundances, with distinctions or relationships having been explored between population sizes, absolute densities and relative densities, between ecological and crude densities, between open and closed populations, between sub-populations, populations and metapopulations, between the abundances of different stage- or age-classes, and between characterizations of the abundances of clonal and non-clonal organisms (e.g. Elton, 1932, 1933; Caughley, 1977; Haila, 1988; Gilpin and Hanski, 1991; McArdle and Gaston, 1993; Gaston, 1994a). This said, it is often not clear to what extent observed patterns in abundances would be altered if they were quantified in a different way, and few studies seek to use more than one measure.

The measurement of the range sizes of species has been much more poorly studied. Nonetheless, it is apparent that different workers tend to employ a variety of methods of measurement (Table 3.1), which inevitably tend to quantify rather different features of the spatial distribution of a species (Gaston, 1991, 1994a,b). A useful distinction can be made between measures of range size which attempt to estimate the extent of occurrence of a species, and those which attempt to estimate its area of occupancy. **Extent of occurrence** is the distance or area between the outermost limits to

Table 3.1 Some measures of the sizes of species geographic ranges, categorized as per Gaston (1994b)

Term	Definition
Linear extent	
Latitudinal extent	Straight-line distance between latitudinally most widely separated occupied sites (Pielou, 1977, 1978; Reaka, 1980; Juliano, 1983; Stevens, 1989; Dennis and Shreeve, 1991; Kouki and Häyrinen, 1991; France, 1992; Rohde *et al.*, 1993)
Longitudinal extent	Straight-line distance between longitudinally most widely separated occupied sites (Reaka, 1980; Juliano, 1983)
Maximum linear extent	Straight-line distance between the two most widely separated occupied sites (Kavanaugh, 1985; Juliano, 1983)
Diagonal distance	$(NS^2 + WE^2)^{1/2}$, where NS and WE are latitudinal and longitudinal extents respectively (Reaka, 1980; Juliano, 1983)
Area within limits	
Biogeographical	Relative size of the biogeographical region in which species are found (Spitzer and Leps, 1988; Thomas, 1991)
Extent	Area within a line, usually drawn by eye, enclosing limits to occurrence (Anderson, 1977, 1984a, 1984b; Glazier, 1980; McAllister *et al.*, 1986; Pagel *et al.*, 1991)
Rectangle	Area of the rectangle defined by the two major perpendicular axes of the species distribution (Stevens, 1986)
'Geometric' circle	Circle defined by the mean value of the radius from the geometric centre of the range to each of the points (Rapoport, 1982)
Minimum circle	Smallest circle containing all occupied localities (Rapoport, 1982)
Minimum convex polygon	Minimum polygon, containing all the localities, in which all internal angles do not exceed 180 degrees (Rapoport, 1982; Juliano, 1983)
Mean propinquity	Overall area about each locality which is enclosed within some function, such as the mean or standard deviation, of the distances between nearest-neighbour localities (Rapoport, 1982)
Numbers of areas occupied	
Geographical areas	Numbers of geographical areas, not usually equal area or at best only approximately so, from which species recorded (Jackson, 1974; Thomas and Mallorie, 1985; McLaughlin, 1992)
Grid	Numbers of quadrats on a grid system from which species recorded (Juliano, 1983; McAllister *et al.*, 1986; Schoener, 1987; Ford, 1990; Pomeroy and Ssekabiira, 1990; Maurer *et al.*, 1991)
Sites	Numbers of sites from which species recorded (McAllister *et al.*, 1986)

the occurrence of a species, and **area of occupancy** is the area over which the species is actually found (Gaston, 1991, 1994a). The former will tend therefore to be the larger, when measured on a truly comparable basis. The magnitude of area of occupancy is strongly dependent on the spatial resolution of the methods used to measure it. Although extent of occurrence measures are frequently applied in ecological studies, measures of area of occupancy are usually more desirable. In particular, the use of extent of occurrence measures will tend to result in potentially artefactual relationships being documented between the size of the range of a species and the breadth or variability of environmental conditions in which it can be found.

Environmental variables tend to show a 'reddened spectrum' (Williamson, 1987), and real relationships between range sizes and environmental tolerance should, as far as is possible, be based on measures of range size which account only for where the species actually occur.

Rather little work has been done on interspecific relationships between range sizes measured in different ways (Gaston, 1994b; but see McAllister *et al.*, 1986; Anderson and Marcus, 1992; Maurer, 1994; Quinn *et al.*, 1996). Work by Quinn *et al.* (1996) on the butterfly and terrestrial mollusc faunas of Britain has documented strong rank correlations between a number of measures of the range sizes (both extents of occurrence and areas of occupancy) of species at this scale, and strong similarities in the identities of those species distinguished as rare on the basis of different measures (defining rare species as the 20% with the smallest estimated range sizes). This suggests that in some circumstances various of these measures are essentially interchangeable. However, the shapes of the relationships between measures can vary markedly (from approximately linear to strongly asymptotic), potentially strongly influencing statements about differences in the absolute range sizes of species, if not their relative range sizes (Figure 3.1).

Depending on the measure of abundance or range size used, different sets of species may be regarded as rare. The extent to which this will affect documented comparisons of the biologies of rare and common species will rest on how different in practice these sets of species are.

3.3.2 Continuous versus discontinuous definitions

Having established a broad definition of rarity based on abundance and range size, distinction needs to be made between operational definitions which treat rarity as continuous and those which treat it as discontinuous. With a continuous definition there are no precise limits to which species are and are not regarded as rare. Extreme rarity and extreme commonness lie at opposite ends of the spectrum of abundances or range sizes, but where commonness ends and rarity begins remains entirely undefined; all species are in some sense rare, but some are rarer than others. Using an

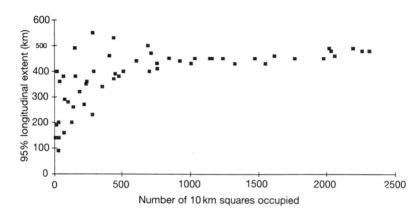

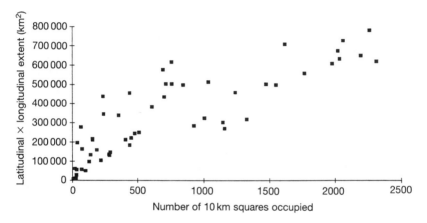

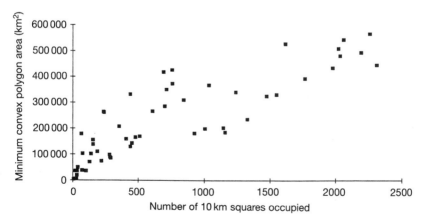

Figure 3.1 Relationships between the numbers of 10 × 10 km squares occupied by resident butterfly species in Britain, and other measures of their range sizes in the region: (a) 95% longitudinal extent; (b) latitudinal × longitudinal extent; (c) minimum convex polygon. (From Quinn *et al.*, 1996.)

operational definition of rarity which is discontinuous, rare species constitute an explicit subset of an assemblage. Species either fall into or out of the rare category.

Although there have been several attempts to determine disjunctions to frequency distributions of the abundances and range sizes of species or repeatable grouping of species of high or low abundance or range size, in general both are continuous variables. Distributions of both tend also to be right-skewed, although at small to moderate spatial scales range sizes may be distributed bimodally (Hanski, 1982; Anderson, 1984a; Schoener, 1987; Lahti *et al.*, 1988; Collins and Glenn, 1991; Gaston, 1994a). The application of a discontinuous definition of rarity necessitates the establishment of an essentially arbitrary cut-off point. Justification of the use of a discontinuous definition of rarity can only therefore be based on pragmatism. Indeed, the use of definitions which are discontinuous seems most widespread where rarity is defined not so much for the purposes of exploring the biology of the phenomenon, as in the creation of lists to highlight species of particular conservation concern. Nonetheless, not infrequently they are applied in an ecological context. Where discontinuous definitions of rarity are used it is seldom clear why the cut-off points applied were chosen, and in some instances actual values of cut-off points are not provided. This makes particular difficulties for comparative studies. For example, it is unclear whether differences in the proportions of species reported by different studies as occurring in each of the categories of Rabinowitz's (1981) scheme, which differentiates seven 'forms of rarity' on the basis of the eight possible combinations of three two-state variables (geographical range – large or small; local population size – large or small; habitat specificity – wide or narrow), reflect real effects or differences in the values of cut-off points (Thomas and Mallorie, 1985; Rabinowitz *et al.*, 1986; Kattan, 1992; Reed, 1992; Mace, 1994). Where discontinuous definitions of rarity are applied the cut-off seems often to be determined on the basis of the proportion of species which it delineates as rare (Gaston, 1994a), this proportion varies widely between studies, with a mode of 20–40% (Figure 3.2).

For the purposes of making comparisons between studies, in the context of a discontinuous definition distinction needs to be made between absolute and relative cut-off points. Using an absolute cut-off point, the abundance or range size value at the cut-off remains constant from one study to another. The position of relative cut-off points is determined on some independent basis, such as the proportion of species which are classified as rare, or a proportion of the summed abundances of all species, and thus the actual value of cut-offs is likely to vary between studies.

Whether continuous or discontinuous definitions of rarity are applied (and, if the latter, what criteria are used) profoundly affects the kinds of questions which can sensibly be asked. For example, using a discontinuous definition with a proportion of species cut-off plainly prevents address to

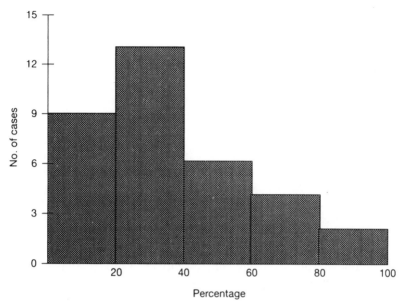

Figure 3.2 Frequency distribution of the numbers of studies defining different percentages of species as rare. (Data from compilation in Gaston, 1994a).

the question of whether high richness communities tend to have more rare species.

3.4 ABUNDANCE × RANGE SIZE

Thus far, abundance and range size have essentially been treated as two separate variables by which species are defined as rare. The question remains as to how independent those two variables are, and how one should treat any interaction.

The interspecific relationship between abundances and range sizes has been widely explored (Gaston, 1994a and references therein; see also Kemp, 1992a,b; Lobo, 1993; Spitzer *et al.*, 1993; Inkinen, 1994; Niemelä and Spence, 1994; Durrer and Schmid-Hempel, 1995). In general, a positive relationship has been documented, with species with high densities or large population sizes tending to have large ranges, and species with low densities or small population sizes tending to have small ranges. The interaction is often rather weak, particularly where numbers of species are large and/or the phylogenetic relatedness of species is low.

A number of hypotheses have been proffered to explain the broad positive relationship, based principally on patterns of resource use and metapopulation dynamics (Brown, 1984; Gaston and Lawton, 1990a; Hanski *et al.*, 1993; Gaston, 1994a). The possible explanations (see Gaston, 1994a, for details) include:

- patterns of aggregation;
- range position;
- vagrancy;
- resource usage (breadth or quantity of);
- metapopulation dynamics;
- artefacts (e.g. sampling effects, averaging over zero values, differences in detectability).

Exploration of these mechanisms continues to be hampered by the difficulty of removing the sampling component from observed interspecific abundance–range-size interactions (Wright, 1991; Hanski *et al.*, 1993; Gaston, 1994a). Species with low local abundances are liable to be recorded from fewer sites regardless of the numbers of sites at which they do in fact occur (a similar problem also complicates analyses of intraspecific relationships between abundance and range size through time; e.g. Marshall and Frank, 1994). Whilst it is acknowledged that this effect can potentially lead to strong, but essentially artefactual, relationships between abundance and range size, its removal necessitates assumptions

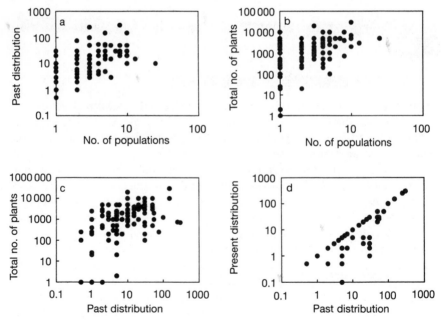

Figure 3.3 Relationships between measures of abundance and range size for threatened Proteaceae in the Cape flora: (a) Past range size (km^2) × number of populations; (b) total number of plants × number of populations; (c) total number of plants × past range size (km^2); (d) present range size (km^2) × past range size (km^2). (Data from Tansley, 1988.)

about the real spatial distributions of the abundances of species which are themselves difficult to validate. Much might be gained through the application of sampling techniques which establish the probability of a species having an abundance below a specified threshold being absent from a site for a given level of sampling effort (e.g. McArdle, 1990). Attention to the temporal dynamics of the interspecific relationship between abundance and range size may also prove valuable.

The distinction between measures of abundance and range size is blurred by methods of quantifying the spatial distributions of species which explicitly weight the occurrences of species in different areas by their abundances there (e.g. Maurer, 1994).

Neither abundance nor range size has any logical precedence as a measure of rarity, and they are often not interchangeable. Thus, it seems reasonable to argue that, where data are available, analyses should consider both variables and their interaction. This approach has been taken further, as mentioned earlier, in the recognition of different forms of rarity on the basis of combinations of abundance and range size. There have been many attempts to recognize, and sometimes label, different forms of rarity (Griggs, 1940; Good, 1948; Mayr, 1963; Drury, 1974; Stebbins, 1978; Terborgh and Winter, 1980; Rabinowitz, 1981; Main, 1984; Cody, 1986; Soulé, 1986; Rabinowitz et al., 1986; Arita et al., 1990; Bawa and Ashton, 1991; McIntyre, 1992). In the context of combinations of abundance and range size, these have largely followed Rabinowitz's (1981) scheme (Rabinowitz et al., 1986; Thomas and Mallorie, 1985; Reed, 1992; McCoy and Mushinsky, 1992). Whilst such an approach has proved useful conceptually, it is perhaps of rather more limited practical value.

3.5 SCALE DEPENDENCE

The abundances and range sizes of species are dynamic both in space and in time. This means that rarity is a scale-dependent concept. Species which may be rare in one area may not be so in another, and species which may be rare over an area of a particular size may not be rare over a larger or smaller area. Likewise, species which are rare in one time period may not be so in another, and species may change their status, as rare or otherwise, when their abundances or range sizes are averaged over periods of differing duration.

The extent to which rare species remain rare in space and in time is arguably a pivotal question in community biology (Strong et al., 1984; Rahel, 1990; Gaston, 1994a). In the spatial context, Schoener (1987, 1990) introduced the terms 'suffusive' and 'diffusive' rarity, to differentiate between species which at a given scale are rare throughout their geographical ranges and those which are rare in some areas but not in others. The same terms can be generalized to other spatial contexts and to temporal ones. Equally, one can consider a more assemblage-based view, expressed in terms of the levels of temporal and spatial concordance of rarity (Rahel, 1990; Gaston, 1994a).

Concordance can be examined at several resolutions:

- individual species tend consistently to fall within the rare category;
- the species composition of the rare category tends to be the same;
- the rank abundances or range sizes of the rare species tend to be the same;
- the rank abundances or range sizes of all the species tend to be the same.

Of these, studies have largely concentrated on concordance in the rank abundances of species in an assemblage. Moreover, to date concordance has received most consideration in a temporal context (e.g. McGowan and Walker, 1985, 1993; Meffe and Minckley, 1987; Johnson and Crowley, 1989; Lawton and Gaston, 1989; Meffe and Sheldon, 1990; Rahel, 1990; Kemp, 1992b; Hansen and Ramm, 1994). Only a few studies have addressed spatial concordance (but see Fager and McGowan, 1963; Grubb *et al.*, 1982; Mitchley and Grubb, 1986; McGowan and Walker, 1993; Watkins and Wilson, 1994).

In addition to spatial and temporal scales, which species are identified as rare depends on both the taxonomic breadth of a study and the way in which species of different status are treated. Evidently, higher taxa may differ in the mean, variance and distribution of the abundances and range sizes of their component species. In at least two published cases, most variance in range sizes is explained at low taxonomic levels (e.g. species), suggesting that here this character appears to be constantly evolving and little affected by phylogeny. Thus, Peat and Fitter (1994) find that, for British angiosperms, 90.5% of variance in number of 10 × 10 km squares occupied in Britain is found at the between-species level, as is 75.6% of the variance in the percentage of European countries in which species are native. Hodgson (1993) finds that more than 50% of the variance in the numbers of 10 × 10 km squares occupied by butterfly species in Britain is found at the generic level.

The effects of species of different status on observed patterns of rarity have been little explored. However, it is clear that in any one area species may occur which have very different potential for persistence there. Crudely, one might perhaps differentiate between residents, migrants and vagrants, and between species with predominantly sink or source population dynamics (Pulliam, 1988). Some groups, such as vagrants, have a differentially high probability of being rare (by continuous or discontinuous definitions). The consequences of their inclusion or exclusion from analyses may therefore be profound. To take one extreme example, up to the end of 1989 the official British and Irish bird list stood at 537 species (there have been a few more recent alterations), of which 258 were termed 'rare' (Whiteman and Millington, 1991). Virtually all of these rare species are vagrants and do not breed in the area. Vagrants are unlikely to be a random subset of species with regard to various other variables (e.g. Gaston *et al.*, 1993), and the inclusion of species of different status may have a very significant effect on observed differences between rare and common species.

3.6 SPECIES SPECIFICITY

The impression is often given that rarity is a species-specific characteristic. It should be plain from previous sections that this can only be so in a very restricted sense. A species may have an abundance or range size which at a global scale is smaller than that of most closely related species. However, it is more useful to view classifications of species as rare or otherwise as entirely contingent on the particular measures of abundance and range size used, spatial and temporal scales, and the taxonomic breadth and other constraints on the assemblage under consideration. Observed interactions between rarity and other variables, such as life history and ecological traits, must therefore always be considered as context dependent. Generalizations from individual studies can only be drawn with caution.

3.7 CONCLUSION

As with so many concepts in ecology, that of rarity can be beguilingly simple. As Buzas *et al.* (1982) warned: 'The list of examples where rare species could lead to erroneous conclusions is limited only by the imagination of the reader.' The variety of definitions used and the wealth of ways in which they can be made operational pose a serious and potentially confusing complication to comparative analyses. One possible solution to this situation would be for future studies to apply a standard definition of rarity alongside any study-specific usage. Such a definition would need to be simple in order to be broadly and readily applicable. In this spirit, Gaston (1994a) advocated the use of a 'quartile' definition of rarity, discriminating the 25% of species in an assemblage with the lowest abundances or smallest range sizes as rare, regardless of how those parameters are measured.

ACKNOWLEDGEMENTS

K.J.G. is a Royal Society University Research Fellow. Bill Kunin kindly commented on the manuscript.

REFERENCES

Anderson, S. (1977) Geographic ranges of North American terrestrial mammals. *American Museum Novitates*, **2629**, 1–15.

Anderson, S. (1984a) Geographic ranges of North American birds. *American Museum Novitates*, **2785**, 1–17.

Anderson, S. (1984b) Areography of North American fishes, amphibians and reptiles. *American Museum Novitates*, **2802**, 1–16.

Anderson, S. and Marcus, L.F. (1992) Areography of Australian tetrapods. *Australian Journal of Zoology*, **40**, 627–651.

Arita, H.T., Robinson, J.G. and Redford, K.H. (1990) Rarity in Neotropical forest mammals and its ecological correlates. *Conservation Biology*, **4**, 181–192.

Batianoff, G.N. and Burgess, R. (1993) Problems in the documentation of rare plants – the Australian experience. *Biodiversity Letters*, **1**, 168–171.

Bawa, K.S. and Ashton, P.S. (1991) Conservation of rare trees in tropical rain forests: a genetic perspective, in *Genetics and Conservation of Rare Plants* (eds D.A. Falk and K.E. Holsinger), Oxford University Press, Oxford, pp. 62–71.

Brown, J.H. (1984) On the relationship between abundance and distribution of species. *American Naturalist*, **124**, 255–279.

Burgman, M.A., Ferson, A. and Akçakaya, H.R. (1993) *Risk Assessment in Conservation Biology*, Chapman & Hall, London.

Buzas, M.A., Koch, C.F., Culver, S.J. and Sohl, N.F. (1982) On the distribution of species occurrence. *Paleobiology*, **8**, 143–150.

Caughley, G. (1977) *Analysis of Vertebrate Populations*, Wiley, New York.

Cody, M.L. (1986) Diversity, rarity, and conservation in Mediterranean-climate regions, in *Conservation Biology: the science of scarcity and diversity* (ed. M.E. Soulé), Sinauer Associates, Sunderland, MA., pp. 122–152.

Collins, S.L. and Glenn, S.M. (1991) Importance of spatial and temporal dynamics in species regional abundance and distribution. *Ecology*, **72**, 654–664.

Dennis, R.L.H. and Shreeve, T.G. (1991) Climatic change and the British butterfly fauna: opportunities and constraints. *Biological Conservation*, **55**, 1–16.

Drury, W.H. (1974) Rare species. *Biological Conservation*, **6**, 162–169.

Drury, W.H. (1980) Rare species of plants. *Rhodora*, **82**, 3–48.

Durrer, S. and Schmid-Hempel, P. (1995) Parasites and the regional distribution of bumblebee species. *Ecography*, **18**, 114–122.

Elton, C. (1932) Territory among wood ants (*Formica rufa* L.) at Picket Hill. *Journal of Animal Ecology*, **1**, 69–76.

Elton, C. (1933) *The Ecology of Animals*, Methuen, London.

Fager, E.W. and McGowan, J.A. (1963) Zooplankton species groups in the North Pacific. *Science*, **140**, 453–460.

Ferrar, A.A. (1989) The role of Red Data Books in conserving biodiversity, in *Biotic Diversity in Southern Africa* (ed. B.J. Huntley), Oxford University Press, Cape Town, pp. 136–147.

Fiedler, P.L. (1986) Concepts of rarity in vascular plant species, with special reference to the genus *Calochortus* Pursh (Liliaceae). *Taxon*, **35**, 502–518.

Fiedler, P.L. and Ahouse, J.J. (1992) Hierarchies of cause: toward an understanding of rarity in vascular plant species, in *Conservation Biology: the theory and practice of nature conservation, preservation and management* (eds P.L. Fiedler and S.K. Jain), Chapman & Hall, London, pp. 23–47.

Ford, H.A. (1990) Relationships between distribution, abundance and foraging specialization in Australian landbirds. *Ornis Scandinavica*, **21**, 133–138.

France, R. (1992) The North American latitudinal gradient in species richness and geographical range of freshwater crayfish and amphipods. *American Naturalist*, **139**, 342–354.

Gaston, K.J. (1991) How large is a species' geographic range? *Oikos*, **61**, 434–438.

Gaston, K.J. (1994a) *Rarity*, Chapman & Hall, London.

Gaston, K.J. (1994b) Measuring geographic range sizes. *Ecography*, **17**, 198–205.

Gaston, K.J. and Lawton, J.H. (1990) Effects of scale and habitat on the relationship between regional distribution and local abundance. *Oikos*, **58**, 329–335.

Gaston, K.J., Blackburn, T.M., Hammond, P.M. and Stork, N.E. (1993) Relationships between abundance and body size – where do tourists fit? *Ecological Entomology*, **18**, 310–314.

Gilpin, M. and Hanski, I. (eds) (1991) *Metapopulation Dynamics: empirical and theoretical investigations*, Academic Press, London.

Glazier, D.S. (1980) Ecological shifts and the evolution of geographically restricted species of North American *Peromyscus* (mice). *Journal of Biogeography*, **7**, 63–83.

Good, R. (1948) *A Geographical Handbook of the Dorset Flora*, The Dorset Natural History and Archaeological Society, Dorchester.

Griggs, R.F. (1940) The ecology of rare plants. *Bulletin of the Torrey Botanical Club*, **67**, 575–594.

Grinnell, J. (1922) The role of the 'accidental'. *The Auk*, **39**, 373–380.

Grubb, P.J., Kelly, D. and Mitchley, J. (1982) The control of relative abundance in communities of herbaceous plants, in *The Plant Community as a Working Mechanism* (ed. E.I. Newman), Blackwell Scientific, Oxford, pp. 79–97.

Haila, Y. (1988) Calculating and miscalculating density: the role of habitat geometry. *Ornis Scandinavica*, **19**, 88–92.

Hansen, M.J. and Ramm, C.W. (1994) Persistence and stability of fish community structure in a southeast New York stream. *American Midland Naturalist*, **132**, 52–67.

Hanski, I. (1982) Dynamics of regional distribution: the core and satellite species hypothesis. *Oikos*, **38**, 210–221.

Hanski, I. (1991) Single-species metapopulation dynamics: concepts, models and observations. *Biological Journal of the Linnean Society*, **42**, 17–38.

Hanski, I., Kouki, J. and Halkka, A. (1993) Three explanations of the positive relationship between distribution and abundance of species, in *Historical and Geographical Determinants of Community Diversity* (eds. R. Ricklefs and D. Schluter), University of Chicago Press, Chicago. pp. 108–116.

Harper, J.L. (1981) The meanings of rarity, in *The Biological Aspects of Rare Plant Conservation* (ed. H. Synge), Wiley, New York, pp. 189–203.

Heywood, V.H. (1988) Rarity: a privilege and a threat, in *Proceedings of the XIV International Congress* (eds W. Greuter and B. Zimmer), Koeltz, Konigstein, Taunus, pp. 277–290.

Hodgson, J.G. (1993) Commonness and rarity in British butterflies. *Journal of Applied Ecology*, **30**, 407–427.

Inkinen, P. (1994) Distribution and abundance in British noctuid moths revisited. *Annales Zoologici Fennici*, **31**, 235–243.

Jackson, J.B.C. (1974) Biogeographic consequences of eurytopy and stenotopy among marine bivalves and their evolutionary significance. *American Naturalist*, **108**, 541–560.

Johnson, D.M. and Crowley, P.H. (1989) A ten year study of the odonate assemblage of Bays Mountain Lake, Tennessee. *Advances in Odonatology*, **4**, 27–43.

Juliano, S.A. (1983) Body size, dispersal ability, and range size in North American species of *Brachinus* (Coleoptera: Carabidae). *Coleopterists Bulletin*, **37**, 232–238.

Kattan, G.H. (1992) Rarity and vulnerability: the birds of the Cordillera Central of Colombia. *Conservation Biology*, **6**, 64–70.

Kavanaugh, D.H. (1985) On wing atrophy in carabid beetles (Coleoptera: Carabidae), with special reference to Nearctic *Nebria*, in *Taxonomy, Phylogeny and Zoogeography of Beetles and Ants* (ed. G.E. Ball), Junk, Dordrecht, pp. 408–431.

Kemp, W.P. (1992a) Rangeland grasshopper (Orthoptera: Acrididae) community structure: a working hypothesis. *Environmental Entomology*, **21**, 461–470.

Kemp, W.P. (1992b) Temporal variations in rangeland grasshopper (Orthoptera: Acrididae) communities in the steppe region of Montana, USA. *Canadian Entomologist*, **124**, 437–450.

Kouki, J. and Häyrinen, U. (1991) On the relationship between distribution and abundance in birds breeding on Finnish mires: the effect of habitat specialization. *Ornis Fennica*, **68**, 170–177.

Kruckeberg, A.R. and Rabinowitz, D. (1985) Biological aspects of endemism in higher plants. *Annual Review of Ecology and Systematics*, **16**, 447–479.

Kunin, W.E. and Gaston, K.J. (1993) The biology of rarity: patterns, causes, and consequences. *Trends in Ecology and Evolution*, **8**, 298–301.

Lahti, T., Kurtto, A. and Väisänen, R.A. (1988) Floristic composition and regional species richness of vascular plants in Finland. *Annales Botanici Fennici*, **25**, 281–291.

Lawton, J.H. and Gaston, K.J. (1989) Temporal patterns in the herbivorous insects of bracken: a test of community predictability. *Journal of Animal Ecology*, **58**, 1021–1034.

Lobo, J.M. (1993) The relationship between distribution and abundance in a dung-beetle community (Col., Scarabaeoidea). *Acta Oecologica*, **14**, 43–55.

Mace, G.M. (1994) Classifying threatened species: means and ends. *Philosophical Transactions of the Royal Society, London, B*, **344**, 91–97.

Main, A. (1982) Rare species: precious or dross?, in *Species at Risk: research in Australia* (eds R.H. Groves and W.D.L. Ride), Springer-Verlag, New York, pp. 163–174.

Main, A.R. (1984) Rare species: problems of conservation. *Search*, **15**, 93–97.

Margules, C. and Usher, M.B. (1981) Criteria used in assessing wildlife conservation potential: a review. *Biological Conservation*, **21**, 79–109.

Marshall, C.T. and Frank, K.T. (1994) Geographic responses of groundfish to variation in abundance: methods of detection and their interpretation. *Canadian Journal of Fisheries and Aquatic Science*, **51**, 808–816.

Maurer, B.A. (1994) *Geographical Population Analysis: tools for the analysis of biodiversity*, Blackwell Scientific, Oxford.

Maurer, B.A., Ford, H.A. and Rapoport, E.H. (1991) Extinction rate, body size, and avifaunal diversity. *Acta XX Congressus Internationalis Ornithologici*, 826–834.

Mayr, E. (1963) *Animal Species and Evolution*, The Belknap Press of Harvard University Press, Cambridge, Massachusetts.

McAllister, D.E., Platania, S.P., Schueler, F.W. *et al.* (1986) Ichthyofaunal patterns on a geographical grid, in *Zoogeography of Freshwater Fishes of North America* (eds C.H. Hocutt and E.D.Wiley), Wiley, New York, pp. 17–51.

McArdle, B.H. (1990) When are rare species not there? *Oikos*, **57**, 276–277.

McArdle, B.H. and Gaston, K.J. (1993) The temporal variability of populations. *Oikos*, **67**, 187–191.

McCoy, E.D. and Mushinsky, H.R. (1992) Rarity of organisms in the sand pine scrub habitat of Florida. *Conservation Biology*, **6**, 537–548.

McGowan, J.A. and Walker, P.W. (1985) Dominance and diversity maintenance in an oceanic ecosystem. *Ecological Monographs*, **55**, 103–118.

McGowan, J.A. and Walker, P.W. (1993) Pelagic diversity patterns, in *Historical and Geographical Determinants of Community Diversity* (eds. R. Ricklefs and D. Schluter), University of Chicago Press, Chicago, pp. 203–214.

McIntyre, S. (1992) Risks associated with the setting of conservation priorities from rare plant species lists. *Biological Conservation*, **60**, 31–37.

McLaughlin, S.P. (1992) Are floristic areas hierarchically arranged? *Journal of Biogeography*, **19**, 21–32.

Meffe, G.K. and Minckley, W.L. (1987) Persistence and stability of fish and invertebrate assemblages in a repeatedly disturbed Sonoran desert stream. *American Midland Naturalist*, **117**, 177–191.

Meffe, G.K. and Sheldon, A.L. (1990) Post-defaunation recovery of fish assemblages in southeastern blackwater streams. *Ecology*, **71**, 657–667.

Mitchley, J. and Grubb, P.J. (1986) Control of relative abundance of perennials in chalk grassland in southern England. I. Constancy of rank order and results of pot- and field-experiments on the role of interference. *Journal of Ecology*, **74**, 1139–1166.

Munton, P. (1987) Concepts of threat to the survival of species used in Red Data books and similar compilations, in *The Road to Extinction* (eds R. Fitter and M. Fitter), IUCN/UNEP, Gland, pp. 72–95.

Niemelä, J.K. and Spence, J.R. (1994) Distribution of forest dwelling carabids (Coleoptera): spatial scale and the concept of communities. *Ecography*, **17**, 166–175.

Pagel, M.P., May, R.M. and Collie, A.R. (1991) Ecological aspects of the geographic distribution and diversity of mammal species. *American Naturalist*, **137**, 791–815.

Peat, H.J. and Fitter, A.H. (1994) Comparative analyses of ecological characteristics of British angiosperms. *Biological Reviews*, **69**, 95–115.

Pielou, E.C. (1977) The latitudinal spans of seaweed species and their patterns of overlap. *Journal of Biogeography*, **4**, 299–311.

Pielou, E.C. (1978) Latitudinal overlap of seaweed species: evidence for quasi-sympatric speciation. *Journal of Biogeography*, **5**, 227–238.

Pomeroy, D. and Ssekabiira, D. (1990) An analysis of the distributions of terrestrial birds in Africa. *African Journal of Ecology*, **28**, 1–13.

Pulliam, H.R. (1988) Sources, sinks, and population regulation. *American Naturalist*, **132**, 652–661.

Quinn, R.M., Gaston, K.J. and Arnold, H.R. (1996) Relative measures of geographic range size: empirical comparisons. *Oecologia*, in press.

Rabinowitz, D. (1981) Seven forms of rarity, in *The Biological Aspects of Rare Plant Conservation* (ed. H. Synge), Wiley, New York, pp. 205–217.

Rabinowitz, D., Cairns, S. and Dillon, T. (1986) Seven forms of rarity and their frequency in the flora of the British Isles, in *Conservation Biology: the Science of Scarcity and Diversity* (ed. M.E. Soulé), Sinauer Associates, Sunderland, Massachusetts, pp. 182–204.

Rahel, F.J. (1990) The hierarchical nature of community persistence: a problem of scale. *American Naturalist*, **136**, 328–344.

Rapoport, E.H. (1982) *Areography: geographical strategies of species*, Pergamon, Oxford.

Reaka, M.L. (1980) Geographic range, life history patterns, and body size in a guild of coral-dwelling mantis shrimps. *Evolution*, **34**, 1019–1030.

Reed, J.M. (1992) A system for ranking conservation priorities for Neotropical migrant birds based on relative susceptibility to extinction, in *Ecology and Conservation of Neotropical Migrant Landbirds* (eds J.M. Hagan III and D.W. Johnston), Smithsonian Institution Press, Washington, pp. 524–536.

Reveal, J.L. (1981) The concepts of rarity and population threats in plant communities, in *Rare Plant Conservation* (eds L.E. Morse and M.S. Henefin), The New York Botanical Garden, Bronx, pp. 41–46.

Rohde, K., Heap, M. and Heap, D. (1993) Rapoport's rule does not apply to marine teleosts and cannot explain latitudinal gradients in species richness. *American Naturalist*, **142**, 1–16.

Schoener, T.W. (1987) The geographical distribution of rarity. *Oecologia* (Berl.), **74**, 161–173.

Schoener, T.W. (1990) The geographical distribution of rarity: misinterpretation of atlas methods affects some empirical conclusions. *Oecologia* (Berl.), **82**, 567–568.

Soulé, M.E. (1986) Patterns of diversity and rarity: their implications for conservation, in *Conservation Biology: the science of scarcity and diversity* (ed. M.E. Soulé), Sinauer Associates, Sunderland, Massachusetts, pp. 117–121.

Spitzer, K. and Leps, J. (1988) Determinants of temporal variation in moth abundance. *Oikos*, **53**, 31–36.

Spitzer, K., Novotny, V., Tonner, M. and Leps, J. (1993) Habitat preferences, distribution and seasonality of the butterflies (Lepidoptera, Papilionoidea) in a montane tropical rain forest, Vietnam. *Journal of Biogeography*, **20**, 109–121.

Stebbins, G.L. (1978) Why are there so many rare plants in California? I. Environmental factors. *Fremontia*, **5**, 6–10.

Stevens, G.C. (1986) Dissection of the species-area relationship among wood-boring insects and their host plants. *American Naturalist*, **128**, 35–46.

Stevens, G.C. (1989) The latitudinal gradient in geographical range: how so many species coexist in the tropics. *American Naturalist*, **133**, 240–256.

Strong, D.R., Lawton, J.H. and Southwood, T.R.E. (1984) *Insects on Plants: Community Patterns and Mechanisms*, Blackwell Scientific, Oxford.

Tansley, S.A. (1988) The status of threatened Proteaceae in the Cape flora, South Africa, and the implications for their conservation. *Biological Conservation*, **43**, 227–239.

Terborgh, J. and Winter, B. (1986) Some causes of extinction, in *Conservation Biology: an evolutionary ecological perspective* (eds M.E. Soulé and B.A. Wilcox), Sinauer Associates, Sunderland, Massachusetts, pp. 119–133.

Thomas, C.D. (1991) Habitat use and geographic ranges of butterflies from the wet lowlands of Costa Rica. *Biological Conservation*, **55**, 269–281.

Thomas, C.D. and Mallorie, H.C. (1985) Rarity, species richness and conservation: butterflies of the Atlas mountains in Morocco. *Biological Conservation*, **33**, 95–117.

Usher, M.B. (1986a) Wildlife conservation evaluation: attributes, criteria and values, in *Wildlife Conservation Evaluation* (ed. M.B. Usher), Chapman & Hall, London, pp. 3–44.

Usher, M.B. (1986b) Insect conservation: the relevance of population and community ecology and of biogeography, in *Proceedings of the 3rd European Congress of Entomology. Part 3* (ed. H.H.W. Velthuis), Nederlandse Entomologische Vereniging, Amsterdam, pp. 387–398.

Watkins, A.J. and Wilson, J.B. (1994) Plant community structure, and its relation to the vertical complexity of communities: dominance/diversity and spatial rank consistency. *Oikos*, **70**, 91–98.

Whiteman, P. and Millington, R. 1991 The British list and rare birds in the eighties. *Birding World*, **3**, 429–434.

Williamson, M.H. (1987) Are communities ever stable? in *Colonisation, Succession and Stability* (eds A.J. Gray, M.J. Crawley and P.J. Edwards), Blackwell Scientific, Oxford, pp. 352–371.

Wright, D.H. (1991) Correlations between incidence and abundance are expected by chance. *Journal of Biogeography*, **18**, 463–466.

4 Who is rare? Artefacts and complexities of rarity determination

Tim M. Blackburn and Kevin J. Gaston

4.1 INTRODUCTION

In order to distinguish those factors or traits that are related to the commonness or rarity of species, one must first somehow define which species are common and which are rare. Most definitions of rarity include consideration either of the geographical range of a species (typically, its extent of occurrence, *sensu* Gaston, 1991), or its abundance within an area, or both together (e.g. Rabinowitz, 1981; Rabinowitz *et al.*, 1986; Gaston, 1994a). Species are generally considered to be rare if their geographical ranges are limited, and/or if their abundances are low. However, what constitutes a low abundance or a limited range can only be determined with reference to the distributions of values of these traits among some relevant set of species. For example, a bird species considered common might attain a density of 100 individuals/km^2, and a bird species considered rare a density of one individual/km^2, but an insect distributed at either of these densities might be considered rare. Thus, rarity is a relative concept (Gaston, 1994a).

Given that rarity is a relative state, it is necessary to compare abundances or spatial distributions in order to decide which of a set of species are rare. This requires that the set of species can, at a minimum, be ranked on whatever criterion is used to assess rarity. Consequently, abundance or range size estimates must be obtained for a number of species. While there is an extensive literature on methods of obtaining comparable estimates of abundances within a species (Caughley, 1977; Seber, 1982), what constitute comparable estimates for abundances of different species is much less clear (Gaston, 1994a; Blackburn and Gaston, 1996).

If interspecific abundance or range size estimates are not comparable,

The Biology of Rarity. Edited by William E. Kunin and Kevin J. Gaston.
Published in 1997 by Chapman & Hall, London. ISBN 0 412 63380 9.

the likelihood arises that species will be misclassified as either rare or common. Common species incorrectly classed as rare may be termed 'pseudo-rare' (type I rarity), whereas rare species incorrectly classed as common exhibit non-apparent (type II) rarity (Gaston, 1994a). These two types of rarity are not independent when the definition of rarity is based on classifying a proportion of a set of species as rare, as the presence of type I rarity in the set means that type II rarity inevitably co-occurs. It is clear that pseudo-rarity and non-apparent rarity are likely if abundance or range size estimates for different species differ in their reliability. However, a more insidious possibility is that they may also occur when abundance or range size estimates for all species in a set are of good or approximately equivalent quality. This possibility could arise if that characteristic of a species that is quantified by an abundance or range size estimate differs between species. The question of how comparable are abundance or range size estimates for different species has received relatively little attention.

This chapter discusses some of the potential problems arising from considerations of rarity based on interspecific comparisons of population densities. As such, we are not concerned with the various methods of counting organisms, but rather with problems that remain when apparently consistent census techniques have been employed. We consider only density estimates for animals. The problems of censusing plant populations are rather different, and probably rather fewer (although seed banks and the recognition of individual organisms may present particular difficulties). While we focus on density, similar problems may equally occur with the other frequently used criterion for judging rarity, geographical range size; some of these have been discussed (see review in Gaston, 1994b). The problems of assessing geographical range size and abundance are not independent, giving rise to the suggestion that, at least in part, the positive correlation frequently observed between these two variables may be artefactual (Wright, 1991; Hanski *et al.*, 1993; Gaston, 1994a).

4.2 COMPARING ANIMAL ABUNDANCES USING DENSITY

When comparing the abundances of animal species, a widely cited piece of dogma is that, on average, large-bodied species tend to be rarer than small-bodied species (e.g. Colinvaux, 1978; Peters, 1991; Dobson and Yu, 1993). In general, a large-bodied organism will require more environmental resources for maintenance and reproduction than will a taxonomically similar organism of smaller body size (Peters, 1983; Calder, 1984). Therefore, equivalent quantities of resources will be able to sustain fewer large-bodied than small-bodied animals, and the former will be rarer. This view is apparently confirmed by studies of the interspecific relationship between body mass and population density in animal taxa (e.g. Damuth, 1981, 1987, 1993; Peters and Wassenberg, 1983; Peters and Raelson, 1984) although the data are equivocal and the subject of considerable debate (Brown and Maurer, 1987; Marquet *et al.*, 1990; Griffiths, 1992; Blackburn *et al.*, 1993, Cotgreave, 1993; Currie, 1993; Blackburn and Lawton, 1994). In such studies, data on the population densities and body masses of a range of animals are compiled from various literature sources. These data compilations generally show negative interspecific relationships between body mass and density, often with high correlation coefficients; small-bodied species are indeed more abundant. We will consider this paradigm repeatedly in the course of this chapter.

Leaving aside the issue that the density of a species is only one possible measure of its rarity (and that large-bodied species are generally more widespread than their small-bodied relatives: Brown and Maurer, 1987, 1989; Gaston, 1994a; Gaston and Blackburn, 1996), what is a comparison of species abundances using a density measure actually comparing? Density is measured as the number of individuals (or some other relevant unit) in a given area. Thus there are two components to any animal density: number of animals and census area. Variation in either component can affect the observed density of a species. While a lot of attention has been focused on methods of determining accurately the actual number of individuals (e.g. Seber, 1982), almost none has been given to the question of the effect of the area over which the density estimate is obtained (but see Haila, 1988). What this area actually is may be relevant in both interspecific and intraspecific comparisons.

4.2.1 Interspecific density comparisons

Consider the interspecific relationship between body mass and abundance described above. One of the earliest studies plotted this for mammalian primary consumers, ranging in size from mice to elephants, and found a strong negative relationship (Damuth, 1981). However, the densities of mice and elephants are unlikely to be censused over areas of similar size. In the data used by Damuth (1987) to explore the same interaction in more detail, there is a strong positive relationship between the body mass of a species and the area over which its density was censused (Figure 4.1; Blackburn and Gaston, 1996). A similar relationship has been shown for mammalian carnivores (Schonewald-Cox *et al.*, 1991). For primary consumers, the area over which a species is censused is a better predictor of its abundance than is its body mass (Figure 4.2; Blackburn and Gaston, 1996).

Many factors determine why an area of a particular size is chosen for study of a given species. Smaller areas will be favoured for practical reasons of relative ease of sampling, delineation of study area, control of disturbance and replication. Conversely, areas must in general be large enough that sufficient numbers of individuals occur within them (species are seldom studied in areas where they are overly difficult to find), and that

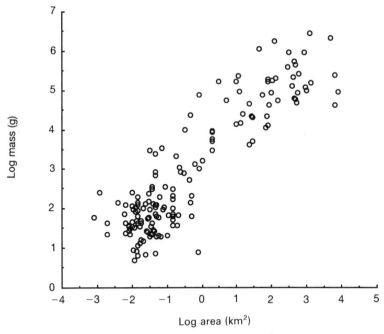

Figure 4.1 Relationship between $\log_{10}$ census area (km^2) and $\log_{10}$ body mass (g) for 160 species of mammalian primary consumer (from Blackburn and Gaston, 1996). Slope = 0.793, r^2 = 0.805, $P < 0.0001$, $n = 160$.

populations are not dominated by transient individuals. Trade-offs between these factors naturally are likely to result in different sized census areas for different sized species (as well as species with different kinds of population dynamics, trophic habits, habitat usage, etc.). It is not immediately obvious that measuring species densities over different sized areas is a problem. However, to what extent are these different sized areas truly comparable? Do they result in measurement of equivalent kinds of ecological density?

There are at least two reasons to doubt that the sizes of areas associated with estimation of the densities of animals of different sizes are strictly comparable. First, it would be necessary that the trade-offs between the various predominantly investigator-oriented, rather than biological, factors determining the size of a study area result in ecologically equivalent areas being used for the measurement of density for both small and large species. This in itself seems exceedingly unlikely. Second, in general, moving from small to large-bodied species, density estimates tend to be derived from increasingly site-oriented rather than species-oriented studies. A high proportion of studies which include density estimates for large mammal species, for example, are studies of the large mammal faunas of particular parks or preserves. Densities derived from site-oriented studies

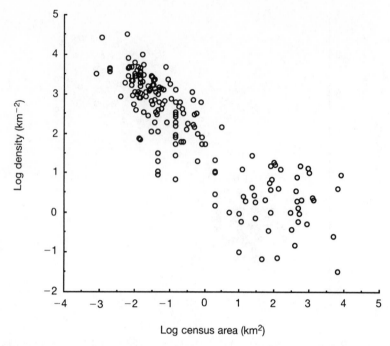

Figure 4.2 Relationship between $\log_{10}$ census area (km^2) and $\log_{10}$ density (km^{-2}) for 160 species of mammalian primary consumer (from Blackburn and Gaston, 1996). Slope = -0.671, $r^2 = 0.736$, $P < 0.0001$, $n = 163$.

typically pay little attention to the patterns of usage of parts of the site by different species. For example, densities may be calculated for a park or preserve for species of a range of body sizes (e.g. dik-dik to elephant), on the basis of precisely the same value of area. Densities of species that use restricted parts of the area hence will be underestimated relative to species using a greater proportion.

The tendency to measure small mammal populations at the scale of trapping grids, and large mammal populations at the scale of nature reserves, suggests the possibility that there may be systematic differences in the relative use made of census areas by animals of different body sizes. Replotting the body-size–abundance relationship for mammalian primary consumers, statistically controlling for the average area over which density was censused for each species, yields a relationship that is still negative, but with both the slope and correlation coefficient much reduced (Blackburn and Gaston, 1996). The clear implications are that different kinds of densities are being measured for large and small animals, and that the interpretation of patterns of abundance from such interspecific comparisons will be confounded by uncertainty as to what is actually being compared.

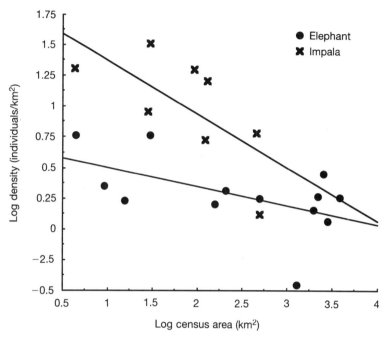

Figure 4.3 Relationship between $\log_{10}$ density and $\log_{10}$ census area in the impala *Aepyceros melampus* (Artiodactyla) (slope $= -0.44$, $r^2 = 0.458$, $P < 0.065$, $n = 8$); and the African elephant *Loxodonta africana* (Proboscidea) (slope $= -0.155$, $r^2 = 0.287$, $P < 0.06$, $n = 13$). Each point is the mean density (individuals/km²) and mean area (km²) in one study (modified from Blackburn and Gaston, 1996).

4.2.2 Intraspecific density comparisons

It is difficult to interpret interspecific body-size–density relationships when a variety of kinds of densities potentially are being measured. The same problem arises for intraspecific density comparisons. Over moderate to large areas, the densities of individual species will tend to decline as area increases, because more sub-areas will be included in which individuals do not occur. Hence, there is no such thing as a single density for a species. Each point in an interspecific plot of density versus area (e.g. Figure 4.2) therefore lies on an intraspecific trajectory. Two examples of such trajectories are given in Figure 4.3, for a species of antelope and a species of elephant. The slopes of these kinds of trajectories are likely to differ between species. The problem for interspecific studies thus becomes one of selecting equivalent densities on which to compare species – which density of *Aepyceros melampus* to compare with which density of *Loxodonta africana* (Figure 4.3). While a logical solution would be to compare densities at some equivalent level of habitat usage – for example the density of a species when 50% of the habitat sampled is utilized by that species – in practice determining such a level is probably impossible for large numbers of species.

One subject area where the problem of density:census area trajectories may have particular relevance is that relating to the examination of the body-size–abundance relationship using phylogenetic methods (Nee *et al.*, 1991; Cotgreave and Harvey, 1991, 1992; Blackburn *et al.*, 1994). Studies using this approach look for possible evolutionary effects on the relationship by testing for patterns within phylogenies and taxonomies, by comparing taxa that share common ancestry. For interspecific analyses of body-size–population density relationships, where variation in body sizes and abundances may span orders of magnitude, it has frequently been argued that the error variance in abundances is so small in relation to the total variance in abundances as to be unlikely to make a significant difference to the observed relationship (e.g. Damuth, 1987; Nee *et al.*, 1991). However, closely related taxa are likely to be similar in size, and frequently similar in abundance. Hence, comparisons between them (as employed by phylogenetic methods) are likely to be affected by error variance in abundance, or by differences in intraspecific density–area trajectories and the positions along those trajectories from which density estimates arise.

4.2.3 Regional versus ecological density

The effect of census area on the density measure obtained has been considered explicitly, though simplistically, in the debate on the interspecific relationship between body size and abundance, in terms of 'regional' and 'ecological' densities (Carrascal and Tellería, 1991; Damuth, 1991; Gaston, 1994a; Gregory and Blackburn, 1995). Put crudely, **ecological density** measures the abundance of species in suitable habitat, whereas **regional density** measures the abundance of species in samples from a given area (Damuth, 1991; Gregory and Blackburn, 1995). Ecological density gives information about the abundance attained by a species within the habitat it occupies, but takes no account of how that area is distributed. Not all of the reported interspecific body-size–abundance relationships are strongly negative; a high proportion show only a weak negative (Juanes, 1986; Brown and Maurer, 1987; Gaston and Lawton, 1988; Cotgreave and Harvey, 1992; Blackburn *et al.*, 1993) or no relationship (Morse *et al.*, 1988; Blackburn *et al.*, 1993). Part of this apparent contradiction has been attributed to different density measures used in different studies: many, though not all, of the studies where weak relationships between density and body size have been reported use regional densities (Juanes, 1986; Gaston and Lawton, 1988; Marquet *et al.*, 1990).

Because of the weak regional density–body-size relationship, it has been suggested that regional density is somehow a less meaningful measure of abundance than is ecological density (exemplified by the names given to the measures: 'ecological' density implies that the measure captures something truly ecological, while the most commonly used alternative name for 'regional' density is 'crude' density), chiefly because it makes no distinction between those parts of the census area where a species does and does not occur. However, the inter- and intraspecific relationships between density and sampling area demonstrate that the regional–ecological distinction is artificial. A whole spectrum of density measures can be produced for a single species depending on the census area used (Haila, 1988). Regional and ecological densities may be towards opposite ends of this spectrum, but the relative positions of these measures on the spectrum are likely to vary from species to species. Whether a species is considered rare or common under a classification scheme based either on regional or on ecological density will depend on how intraspecific density trajectories (e.g. Figure 4.3) overlay at the point or points on the area axis selected for density comparisons. The fundamental question of what constitutes a species density, given a wide range of possible answers, is not addressed by either measure.

4.2.4 Comparing densities in habitat patches

The examples discussed above showed different negative relationships between density and area in different species in different study areas of vastly different sizes. A reasonable conclusion might be that it is inappropriate to compare the density of mice on a trapping plot with the density of elephants in a game reserve, because the species in question are likely to make markedly different relative use of these different areas.

Unfortunately, the solution to the problem of area and density is not simply to avoid comparisons at greatly differing scales, or to avoid selecting obviously inappropriate areas for comparison. Similar species may show markedly different density–area trajectories when their abundances are measured at the same sites. For example, Nilsson (1986) examined the breeding densities of various species of water birds on lakes of differing sizes in south Sweden. He found that, while several water bird species breed at higher densities on lakes of small area than they do on lakes of large area, other species are absent from small lakes entirely, and some of those appear to breed at higher densities on large lakes than on small. The breeding density of only one species seemed unaffected by lake size (Nilsson, 1986).

For the water bird data, whether a species is defined as rare or common could depend on the lake or set of lakes on which population censuses are carried out, because different species make different uses of lakes of a given size. In general, the estimate of density obtained for different species can depend critically on characteristics of the habitat patch (or patches) in which abundance is censused. This can be either because of differences in the way species use areas (as in the water birds), or despite them. It is easy to imagine an example of two bird species using exactly the same proportion of habitat in an area of woodland, but with one species preferring forest edges and the other only occurring away from them. If forest edges are easier to survey, the species with a preference for edge areas might appear more abundant, regardless of the actual numbers of individuals present overall. Even standardizing the census area and the proportion of a given area used by a species may not be sufficient to give a true impression of which species are rare and which common.

4.2.5 Density at the lowest level – home range size

Even given a reliable estimate of the number of individual animals in an area, the measurement of density is clearly problematic. The area over which density is sampled affects the answer obtained. Larger areas are likely to include a higher proportion of space unused by the species being censused. How a given area is used is likely to be species-specific. To compare the densities of two or more species, equivalent kinds of densities are required, in turn requiring density measurement over areas which are ecologically equivalent for the species in question.

One possible solution to these problems is to compare densities at the lowest level for each species: that is, the density of an average individual animal, or a function of the mean size of the home range of a species. Home range sizes have been used to estimate species densities in the past, particularly for taxa in which individuals range widely and in which overlap in their use of space is small (e.g. Newton, 1979). However, such attempts have tended to be based on some of the less sophisticated measures of home range size (e.g. minimum convex polygons), rather than those based on utilization functions (probability density functions of location; Worton, 1987). These latter approaches may provide a means of generating comparability in density measures between species through some equivalence in the ways in which they use space. The problems of unused areas in a density estimate are reduced, because home range size is defined by individual usage of an area.

If we sound uncharacteristically optimistic at this point, it is worth noting that comparisons of species densities based on home range sizes will have additional problems to overcome. In particular, consideration of situations where home ranges overlap, and where there are significant gaps between home ranges, will be necessary before comparisons can be treated with confidence (an integration of home range size with number of individuals is likely to be required). Also, temporal aspects of density determination may be more significant problems when comparisons are at low levels (Chapter 3). Nevertheless, at present, home range size seems to offer the best hope for eliminating the problems of interspecific differences in the use of areas.

4.2.6 Dimensionality

We have shown that categorizations of species abundance based on comparisons of densities are likely to be flawed, because there is an interaction between the area over which density is censused and the density value obtained. So far, we have assumed that there is such a thing as an absolute area, that area is invariant. However, area has fractal properties. The perceived size of an area will depend on the size of ruler used to measure it. For example, the area of a game reserve perceived by a scientist from a map may be very different to that perceived by an antelope using that area, and different again from that perceived by a mouse. Thus, the density–area relationship described here may itself be dependent on the fractal dimension of habitat area. (For some other possible ecological implications of fractal habitat dimensions, see Morse *et al.*, 1985 and Shorrocks *et al.*, 1991.)

Moreover, we have simplistically assumed that most species perceive habitat in two dimensions. For many species, a three-dimensional view is certainly more appropriate. Indeed, this has been suggested as a possible reason for the general weakness of interspecific body-size–density relationships in birds (Juanes, 1986; Cotgreave and Harvey, 1992); birds are perhaps more likely to perceive the environment in three dimensions. How the number of dimensions in which species utilize the environment might affect the observed interspecific relationship between body size and density (e.g. Damuth, 1981, 1987), where densities of species like squirrels and primates are compared with densities of species like antelopes and elephants, can only be guessed. The body-size–density relationship in birds was not improved by restricting analysis to species that are either flightless or make only limited use of their ability to fly (Cotgreave and Harvey, 1992).

4.3 CONCLUSION

There is an increasing number of examples of ecological relationships where observed patterns are likely to be seriously compromised by methodological problems, including relationships to the temporal variability of populations (McArdle *et al.*, 1990; Gaston and McArdle, 1994), body-size distributions (Blackburn and Gaston, 1994a, b), and abundance–geographical range size relationships (Wright, 1991; Hanski *et al.*, 1993; Gaston, 1994a). To this list must now be added relationships involving population density. The problems that we have highlighted in the preceding sections suggest that there are almost as many kinds of density as there are studies measuring it. Thus, defining what is meant by a species 'density' will be extremely problematic, and additionally specifying criteria whereby species densities can be compared will be at least as difficult.

Interspecific comparisons using density are the most likely to suffer from the problems we have discussed, because they compound the problems of both intra- and interspecific comparisons. Hence, studies that attempt to find correlates of species densities, or that rank species on a density measure, are particularly likely to be affected; any classification of species into different categories of rarity using density is clearly included in this set. Given that conservation strategies and other decisions affecting the survival probability of species or populations will be based on abundance information, we advocate caution where that basis is comparative density information. Haila (1988) notes that 'different "densities" are answers to different questions'. When the question is that of rarity, it is especially important that it is the same question for all species.

REFERENCES

Blackburn, T.M. and Gaston, K.J. (1994a) Animal body size distributions change as more species are described. *Proceedings of the Royal Society of London, B,* **257**, 293–297.

Blackburn, T.M. and Gaston, K.J. (1994b) Animal body size distributions: patterns, mechanisms and implications. *Trends in Ecology and Evolution*, **9**, 471–474.

Blackburn, T.M. and Gaston, K.J. (1996) Abundance–body size relationships: the area you census tells you more. *Oikos*, **75**, 303–309.

Blackburn, T.M. and Lawton, J.H. (1994) Population abundance and body size in animal assemblages. *Philosophical Transactions of the Royal Society of London, B*, **343**, 33–39.

Blackburn, T.M., Brown, V.K., Doube, B.M. *et al.* (1993) The relationship between body size and abundance in natural animal assemblages. *Journal of Animal Ecology*, **62**, 519–528.

Blackburn, T.M., Gates, S., Lawton, J.H. and Greenwood, J.J.D. (1994) Relations between body size, abundance and taxonomy of birds wintering in Britain and Ireland. *Philosophical Transactions of the Royal Society of London, B*, **343**, 135–144.

Brown, J.H. and Maurer, B.A. (1987) Evolution of species assemblages: effects of energetic constraints and species dynamics on the diversification of the American avifauna. *American Naturalist*, **130**, 1–17.

Brown, J.H. and Maurer, B.A. (1989) Macroecology: the division of food and space among species on continents. *Science*, **243**, 1145–1150.

Calder, W.A. (1984). *Size, Function and Life History*, Harvard University Press, Cambridge, Mass.

Carrascal, L.M. and Tellería, J.L. (1991) Bird size and density: a regional approach. *American Naturalist*, **138**, 777–784.

Caughley, G. (1977) *Analysis of Vertebrate Populations*, Wiley, New York.

Colinvaux, P. (1978) *Why Big Fierce Animals Are Rare*, Princeton University Press, Princeton, New Jersey.

Cotgreave, P. (1993) The relationship between body size and abundance in animals. *Trends in Ecology and Evolution*, **8**, 244–248.

Cotgreave, P. and Harvey, P.H. (1991) Bird community structure. *Nature*, **353**, 123.

Cotgreave, P. and Harvey, P.H. (1992) Relationships between body size, abundance and phylogeny in bird communities. *Functional Ecology*, **6**, 248–256.

Currie, D.J. (1993). What shape is the relationship between body size and population density? *Oikos*, **66**, 353–358.

Damuth, J. (1981) Population density and body size in mammals. *Nature*, **290**, 699–700.

Damuth, J. (1987) Interspecific allometry of population density in mammals and other animals: the independence of body mass and population energy use. *Biological Journal of the Linnean Society*, **31**, 193–246.

Damuth, J. (1991) Of size and abundance. *Nature*, **351**, 268–269.

Damuth, J. (1993) Cope's rule, the island rule and the scaling of mammalian population density. *Nature*, **365**, 748–750.

Dobson, F.S. and Yu, J. (1993) Rarity in Neotropical forest mammals revisited. *Conservation Biology*, **7**, 586–591.

Gaston, K.J. (1991) How large is a species' geographic range? *Oikos*, **61**, 434–437.

Gaston, K.J. (1994a) *Rarity*. Chapman & Hall, London.

Gaston, K.J. (1994b) Measuring geographic range sizes. *Ecography*, **17**, 198–205.

Gaston, K.J. and Blackburn, T.M. (1996) Some conservation implications of geographic range size–body size relationships. *Conservation Biology*, **10**, 638–646.

Gaston, K.J. and Lawton, J.H. (1988) Patterns in body size, population dynamics, and regional distribution of bracken herbivores. *American Naturalist*, **132**, 662–680.

Gaston, K.J. and McArdle, B.H. (1994) The temporal variability of animal abundances: measures, methods and patterns. *Philosophical Transactions of the Royal Society of London, B*, **345**, 335–358.

Gregory, R.D. and Blackburn, T.M. (1995) Abundance and body size in British birds: reconciling regional and ecological densities. *Oikos*, **72**, 151–154.

Griffiths, D. (1992) Size, abundance, and energy use in communities. *Journal of Animal Ecology*, **61**, 307–315.

Haila, Y. (1988) Calculating and miscalculating density: the role of habitat geometry. *Ornis Scandinavica*, **19**, 88–92.

Hanski, I., Kouki, J. and Halkka, A. (1993) Three explanations of the positive relationship between distribution and abundance of species, in *Species Diversity in Ecological Communities: historical and geographical perspectives*, (eds R.E. Ricklefs and D. Schluter), University of Chicago Press, Chicago, pp. 108–116.

Juanes, F. (1986) Population density and body size in birds. *American Naturalist*, **128**, 921–929.

Marquet, P.A., Navarette, S.A. and Castilla, J.C. (1990) Scaling population density to body size in rocky intertidal communities. *Science*, **250**, 1125–1127.

McArdle, B.H., Gaston, K.J. and Lawton, J.H. (1990) Variation in the size of animal populations: patterns, problems and artefacts. *Journal of Animal Ecology*, **59**, 439–454.

Morse, D.R., Lawton, J.H., Dodson, M.M. and Williamson, M.H. (1985) Fractal dimensions of vegetation and the distribution of arthropod body lengths. *Nature*, **314**, 731–733.

Morse, D.R., Stork, N.E. and Lawton, J.H. (1988) Species number, species abundance and body length relationships of arboreal beetles in Bornean lowland rain forest trees. *Ecological Entomology*, **13**, 25–37.

Nee, S., Read, A.F., Greenwood, J.J.D. and Harvey, P.H. (1991) The relationship between abundance and body size in British birds. *Nature*, **351**, 312–313.

Newton, I. (1979) *Population ecology of raptors*. Buteo Books, Vermillion, S. Dakota.

Nilsson, S. G. (1986) Are bird communities in small biotype patches random samples from communities in large patches? *Biological Conservation*, **38**, 179–204.

Peters, R.H. (1983) *The Ecological Implications of Body Size*, Cambridge University Press, Cambridge.

Peters, R.H. (1991). *A Critique for Ecology*, Cambridge University Press, Cambridge.

Peters, R.H. and Raelson, J.V. (1984) Relations between individual size and mammalian population density. *American Naturalist*, **124**, 498–517.

Peters, R.H. and Wassenberg, K. (1983) The effect of body size on animal abundance. *Oecologia*, **60**, 89–96.

Rabinowitz, D. (1981) Seven forms of rarity, in *The Biological Aspects of Rare Plant Conservation* (ed. H. Synge), J. Wiley and Sons Ltd., New York, pp. 205–217.

Rabinowitz, D., Cairns, S. and Dillon, T. (1986) Seven forms of rarity and their frequency in the flora of the British Isles, in *Conservation Biology: the science of scarcity and diversity* (ed. M.E. Soulé), Sinauer Associates, Sunderland, Massachusetts, pp. 182–204.

Schonewald-Cox, C., Azari, R. and Blume, S. (1991) Scale, variable density, and conservation planning for mammalian carnivores. *Conservation Biology*, **5**, 491–495.

Seber, G.A.F. (1982) *The Estimation of Animal Abundance and Related Parameters*, 2nd edn, MacMillan, New York.

Shorrocks, B., Marsters, J., Ward, I. and Evennett, P.J. (1991) The fractal dimension of lichens and the distribution of arthropod body lengths. *Functional Ecology*, **5**, 457–460.

Worton, B.J. (1987) A review of models of home range for animal movement. *Ecological Modelling*, **38**, 277–298.

Wright, D.H. (1991) Correlations between incidence and abundance are expected by chance. *Journal of Biogeography*, **18**, 463–466.

PART TWO: MECHANISMS CREATING RARE–COMMON DIFFERENCES

5 Who gets the short bits of the broken stick?

Michael L. Rosenzweig and Mark V. Lomolino

5.1 INTRODUCTION

'Extinction is simply limitation of population density carried to the extreme,' wrote Stanley (1979). So, we concern ourselves with rarity because it is the precursor of extinction. If we knew what predisposes a species toward rarity, we might be able to forestall some extinctions. At least, we might find ways to slow extinction down.

But perhaps nothing predisposes toward rarity. Perhaps instead, species abundances are simply the chance outcomes of stochastic processes that follow statistical rules. Preston's canonical log-normal distribution is one such rule, and MacArthur's broken stick is another (May, 1975). According to this view, the possession of any relative abundance is a mere accident. Nature governs only the overall distribution of abundances. Trying to predict which species will get rare and go extinct is rather like trying to predict which molecules of water will boil away as you heat your morning coffee.

Are species like water molecules? Are their abundances like the distribution of lengths of straw in a lottery? Copepod species abundances in the North Pacific and the South Pacific gyres suggest a clear answer. Rare species really are different. They stayed rare in samples widely separated in ecological time and space (McGowan and Walker, 1985, 1993). Rare species even remained rare during and after a major climatic perturbation that lasted for a year and doubled average copepod densities. Such data hint that at least some aspects of rarity must not be so capricious as the fate of a water molecule.

In this chapter we will assume that the copepods are pointing in the right direction. Let us see how far we can get using that assumption. We will seek ecological properties of species that may be causally associated with rarity. We do not offer this chapter as a definitive review of each of its many parts; that would be a book-length effort. Nor do we expect that our

The Biology of Rarity. Edited by William E. Kunin and Kevin J. Gaston.
Published in 1997 by Chapman & Hall, London. ISBN 0 412 63380 9.

categorization will stand without considerable modification by later ecologists. Instead, we hope that you will see it as two ecologists' view of the forest, as a searching for the kinds of questions that will make us better conservation biologists.

Our definition of **rarity** is: low total number of individuals in a species – in other words, low abundance. 'Density' means something else (although Stanley does use it to mean 'abundance' in his epigraph, above). **Density** means 'number of individuals per unit of spatial sample at one instant'. Thus a broad-ranged species may be sparse but not rare, and a dense species may (theoretically at least) be so restricted in distribution as to be rare and endangered. Nevertheless, low density and narrow geographical distribution constitute the elements of rarity.

At first, it seemed to us that three broad sorts of traits may predispose a species toward rarity:

- Some species need certain restricted habitats.
- Some have interactions with competitors or predators that severely limit their population.
- Some species occupy a high position in their food web.

But these three general categories are probably not useful enough. Consider predation, for example. Exploiters may severely limit the populations of their victims (Crawley, 1992). Ecologists often take advantage of this fact to control pests and weeds. But, except for viruses, virtually every species has its enemies. Why aren't they all rare? And how would we recognize rarity – surely it is a relative concept – if all victims were similarly limited by their predators? Can we discern certain kinds of exploitative interactions most likely to cause rarity? Or is the result a spin of the roulette wheel?

Our questions led us to abandon the three most general causes of rarity and seek a slightly more refined set that may be more useful. We suggest the following list of eight candidate traits, each of which may predispose to rarity or even produce it.

- Some species are intolerant.
- Some species occupy a high trophic level (we will suggest this category is wrong).
- Some species occupy an extreme niche in a guild of species with distinct preferences.
- Some species depend on habitats that become rare as a result of climate changes.
- Some species belong to taxa of high habitat selectivity.
- Some species are the products of centrifugal speciation.
- Some species are governed by chaotic interactive dynamics.
- Some species are vulnerable to new competitive or exploitative interactions.

Except for the trophic level attribute, which occurs in both lists, you cannot map the eight attributes of the second list into the first (i.e. habitat restriction, negative interactions and high trophic level). For instance, you will see that intolerant species often need restricted habitats, but would use a wider variety were it not for their interactions with other species. Species suffering from a climate change might also be quite common were they not restricted to a subset of habitats by negative interactions. Nor can we achieve much by merely dividing negative interactions into exploitation and competition: both extreme niches and chaotic dynamics may arise from a combination of competition and exploitation. Finally, neither climate change nor centrifugal speciation are even contained in the list of three. Thus, we have not tried to construct the list of eight in order to subdivide the list of three.

Nor have we done it to outbid Rabinowitz's (1981) famous list. That list focuses on symptoms. Ours tries to identify processes and mechanisms. We believe that looking in the direction of causes (rather than symptoms) may improve us as conservation biologists. Hence, a search for causation should expand the discourse in a useful direction. Let us explain, evaluate and discuss each of the eight candidates in turn.

5.2 A CANDIDATE LIST OF ECOLOGICAL CAUSES OF RARITY

5.2.1 Rarity and intolerance – the thick bits

Hanski (1982; Hanski et al., 1993) has led us to understand that rarity and narrow geographical range usually go together. We believe this strongly suggests that some regular process helps to determine rarity. We further believe that this process is competitive coexistence by means of tolerance diversity. What is meant by such coexistence?

Field investigations often reveal a pattern of fundamental niches called **included niches** (Colwell and Fuentes, 1975; Chase and Belovsky, 1993). Species in a guild have niche positions that all include one particular type of habitat, but some species can use more than that one, and others cannot. The latter are the intolerant species. Some species, in fact, can use many, many more habitat types. These are the most tolerant.

One mechanism – **shared preference habitat allocation** – may account for most if not all included-niche patterns. This mechanism obtains if the habitat that every species can use is also the best habitat for all species, i.e. if tests of ability show that all species perform best in it (Abramsky et al., 1990; Rosenzweig, 1974, 1979, 1991). Thus the shared habitat is also the shared preference. The mechanism also depends on a trade-off between niche breadth and the ability to take best advantage of the preferred habitat: species without the ability to use most habitats must be able thereby to dominate the preferred habitat and survive.

The literature abounds with proven and likely examples of coexistence by tolerance variation. Some are classical. *Balanus balanoides* cannot tolerate the highest part of the intertidal zone, but dominates *Chthamalus stellatus* lower down by growing faster (Connell, 1961). *Orconectes virilis* suffocate in still, muddy freshwater ponds because of their high metabolism, but they take advantage of that higher level of activity to drive out *Orconectes immunis* from the richly oxygenated rocky shallows that both species prefer (Bovbjerg, 1970). *Lampornis clemenciae* lives only in the rich riparian canyon bottoms preferred by it, and by the two coexisting species of hummingbird, *Eugenes fulgens* and *Archilochus alexandrii*. It uses its larger body and inefficient but highly manoeuvrable wing design to force them to the drier hillsides (Pimm *et al.*, 1985). Acanthocephalans and tapeworms both prefer sites in a rat's gut about 20% of the length away from the pylorus. That distance provides a habitat where much of a rat's food is digested but little has already been absorbed – gut-parasite heaven. Even under intense interspecific competition, the acanthocephalans do not abandon it, but they force the tapeworms further down the gut when the two occur together (Holmes, 1961).

Notice that coexistence based on variation in tolerance is not the same as interference competition (Brown, 1989). Although the intolerant species may use interference competition to dominate the preferred habitat, it has alternatives. For example, intolerant species may find preferred patches quicker than tolerant ones do. Or they may retreat metabolically in seasons when only secondary patches occur.

On the other hand, tolerance variation and included niches have a great deal in common. Included-niche communities have a number of species that share a modal niche position, but their niche breadths vary significantly so that successively broader species fully contain the narrower niches. If these niches are the fundamental niches and there is a trade-off between breadth and competitive ability at the mode, then the included-niche pattern is the tolerance variation pattern. Some have thought the included pattern rare (see references in Chase and Belovsky, 1993), but data suggest it may be seen as the most common if its relationship to tolerance variation is appreciated (Rosenzweig, 1991). Tolerance variation is the mechanism that often results in the pattern of included niches.

What does variation in tolerance have to do with rarity? Often the preferred habitat is an extreme one, i.e. extremely productive. We expect extremes of any sort to be relatively scarce and restricted, and this is often (but not always) true of extremely productive habitats. When it is, the intolerant species will be concomitantly rare and restricted. For example, *Eugenes fulgens* and *Archilochus alexandrii* are much more common and widespread than their intolerant competitor *Lampornis clemenciae*. Because intolerant species often require scarce but rich habitats, we say that they tend to receive the short, thick bits of the broken stick.

Another source of population-size differences is the expense of producing and maintaining one individual. Successful intolerant species are often larger than tolerants, or have higher metabolisms, or higher growth rates, or produce expensive allelochemicals to use in interference competition. Such differences may not be enough to cause real rarity, but they will augment any disparity caused by habitat scarcity.

The preferred habitat is not always so scarce. For example, the resource-poor high intertidal zone cannot exist without the resource-rich low intertidal. So, despite its being an intolerant species unable to live in the high intertidal zone, we do not expect *Balanus balanoides* to be rare (and it is not).

The point is not the exceptions. It is the bias that comes from intolerance because the preferred habitat is often scarce. Intolerant species are more likely to be rare for several reasons:

- They are adapted to few habitats, often only the richest.
- Those few habitats are more likely to be scarce than poorer ones.
- The cost of dominating the best habitats reduces the efficiency with which intolerant species convert energy and resources into intolerant individuals. The result is fewer individuals, even in the habitats that support them.

5.2.2 Rarity and trophic level – the high bits

High trophic level has long been suspected to cause rarity, but evidence suggests that this hypothesis is wrong. The trophic level hypothesis comes from noting that free energy declines as it travels up the food chain. Each succeeding trophic level has only about 10–15% of the energy available to the preceding one. So, it would seem, species at higher trophic levels should have fewer individuals, other things being equal.

Aye, there's the rub. Other things are not equal. Body sizes of species often change along a trophic chain, and diversity certainly does. We will argue that changes in diversity largely cancel out the energetic effects of trophic position, leaving mostly the effects of body size. We doubt that trophic level affects body size consistently enough to produce a trend.

Schoenly *et al.* (1991) collected the reports of insect food webs in many communities. In these reports, *ca.* 80% of the nodes were species. Using their data and definition of trophic level (i.e. the maximum, the highest trophic level at which a species is known to feed), Rosenzweig (1995) grouped the nodes by trophic level and found that animal diversity in insect food webs declines semi-logarithmically with trophic level (Figure 5.1). But the decline seems too gentle to even out the average population size across trophic levels, because it would only do so if 34.4% of free energy were transferred from one level to the next. Instead, we know that only about 10–15% gets transferred. Thus, higher trophic levels seem to have more species per unit of free energy than do lower ones. If that is true, then

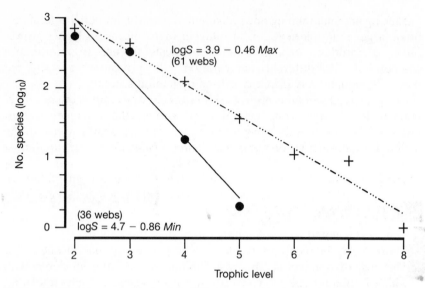

Figure 5.1 Number of species at different trophic levels in insect-dominated communities. The shallower regression line (· · · –) is the best estimate of the data when species are assigned to their maximum trophic level (pluses). The steeper line (solid) is the best estimate when species are assigned to their minimum trophic level (large dots). (Redrawn from Rosenzweig, 1995.)

so is conventional wisdom: species of higher trophic levels should be rarer on average. But is it true?

The answer depends on whether the maximum trophic level properly indicates the energy received by a species. It does not. Maximum trophic level overestimates the number of links that energy traverses to get to a species. Worse, that bias increases with trophic level: an animal recorded at maximum trophic level two really resides at level two, but one recorded at level six probably gets almost all of its food from lower levels. Thus, the slope generated by the study of maximum trophic levels is too gentle. Is there a reasonable way to correct it? Yodzis (1993) suggests that 'minimum trophic level' yields a better estimate of the average number of links that energy passes to reach a species. In fact, since *ca.* 85–90% of free energy dissipates owing to passage through one link, Yodzis's opinion ought to be correct. If energy declines by about one order of magnitude at each passage, a species feeding indiscriminately on, say, three trophic levels will get the equivalent of 0.01 food unit from the highest for every 1.0 unit from the lowest. Its average trophic level would be $n + (0.12/1.11) = n + 0.108$, where n is the lowest level it feeds on. This is certainly much closer to n, the minimum trophic level, than it is to $n + 2$, which is the maximum trophic level.

Rosenzweig (1995) re-examined most of the reports cited by Schoenly *et al.* (1991) and determined the minimum trophic level wherever possible.

This resulted in a slope consistent with a loss of about 86.2% of free energy at each trophic transfer (Figure 5.1). Such a slope is steep enough to maintain the same average energy per animal species regardless of trophic level. Thus, we cannot attribute rarity to trophic level.

Among vertebrates (Vézina, 1985) and invertebrates (Warren and Lawton, 1987) individual victims are eaten by larger predators (Cohen *et al.*, 1993). This suggests that body size rises with trophic level. If it does, higher trophic levels should support fewer individuals from each kJ/m^2 per year. It is the number of individuals in the standing biomass – not the energy flow, *per se* – that measures abundance. So, species at higher trophic levels might tend to be rarer. But parasitic taxa probably show a reverse body-size trend and we ecologists hardly ever include them in our food web studies (Warren and Lawton, 1987). Perhaps parasitic taxa would neutralize the body-size trend entirely were we to sample enemy–victim size ratios without bias? In any case, we believe that the presumed effect of trophic level on rarity probably stems from a perspective too narrowly focused on fierce predators, especially vertebrates. If body size does alter a species' chances to be rare, then it is body size that should be named and studied. Trophic level may play no part at all.

5.2.3 Rarity and distinct preferences – the end bits

Ecological organization of species into different niches does not always depend on shared–preference habitat selection. In fact, the first theories of community ecology dealt only with a model in which each species has distinct preferences. A common example is the case of a guild of consumers whose species have different body sizes and thus specialize on resources of different size. Mammalian carnivores seem to do that (Rosenzweig, 1966). So do certain raptors (Storer, 1966), granivorous ants (Davidson, 1977), tiger beetles (Pearson and Mury, 1979) and many, many other groups of species.

Each bauplan of animal appears to have an optimal body size (Roff, 1981; Brown *et al.*, 1993). In the absence of competitors, it will evolve to be that size. But new species will have a different optimum, larger or smaller than the single species by itself. Thus, a guild of co-occurring species consists of one species that fills the dream-niche of the bauplan, and other species that natural selection has forced toward opposite extremes, large or small.

The body-size distribution of the resources of such a group of species (or consumer guild) may well provide one reason to believe that medium-sized consumers often fill the ideal niche of a bauplan. It would be unusual if those resources did not have an abundance that peaks over some intermediate resource size. Moreover, such distributions tend to be skewed toward the smaller resources so that we might expect mostly the extremely large-sized resources to be rare. In turn, the extremely large body sizes of the guild of consumers should also tend to be rare. Wolves should be rarer than foxes, goshawks rarer than sharp-shinned hawks, peregrines rarer than kestrels, etc. Perhaps this is the real source of the conviction that high trophic levels often produce rare species. Many rare species do seem to be large predators in a distinct-preference body-size guild.

The endangered American burying beetle, *Nicrophorus americanus*, is one case of a once abundant and widespread species that has become very rare because it is the largest member of its guild (Lomolino *et al.*, 1995). During the past 50 years, its range has collapsed. Now it survives only as a few remnant populations on the periphery of its former range (Figure 5.2).

Burying beetles feed and breed on carcasses of vertebrates, primarily mammals and birds. *N. americanus* requires larger carcasses than others of its guild (50–300 g *vs.* < 50 g; Kozol *et al.*, 1988). Consequently, even in their remnant strongholds, *N. americanus* tends to be rare – much rarer than the smaller burying beetles. The most abundant syntopic burying beetle species is typically two orders of magnitude more dense (Lomolino *et al.*, 1995).

We should add that habitat as well as extreme body size may contribute to rarity in this case: *N. americanus* may require habitats with loose, deep soils and substantial accumulations of litter because larger carcasses are more difficult to bury. Such substrates characterize mature forests, and Anderson (1982) suggested that deforestation in the USA led to the decline of *N. americanus*. Both *N. germanicus* of Europe (now presumed

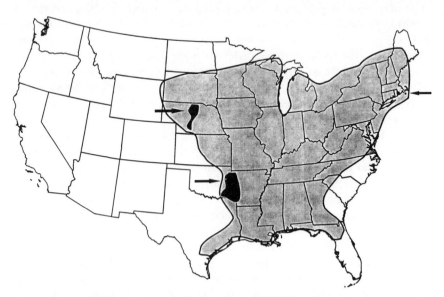

Figure 5.2 Collapse of the geographical range of the American burying beetle, *Nicrophorus americanus*. Historic range (1880–1940) indicated by shading; extant populations (1990 to present) indicated in black with arrows. (Redrawn from Lomolino *et al.*, 1995.)

extinct) and *N. concolor* of China and Japan (whose status is not known to us) are or were forest specialists and the largest local members of their guild.

Recent studies (Lomolino *et al.*, 1995, and in preparation) support Anderson's hypothesis. In the Ouchita National Forest (southeastern Oklahoma) *N. americanus* are seven times as common in mature forests as in second growth, and 70 times as common in mature forests as in clearcuts ($\chi^2 = 109.4$; $P < 10^{-3}$). Breeding success was also greater in mature forests: of 27 pairs of *N. americanus* put on carcasses in grassy habitats, only 15 succeeded in burying the carcass and rearing young, but 21 of 22 did so on carcasses in forested sites. Moreover, each carcass produced more young in a forested site (14.77 *vs.* 9.79; $t = 2.18$, $P < 0.05$).

In requiring large resources and being relatively uncommon, *N. americanus* typifies large guild members (Rosenzweig, 1966; Ashmole, 1968; Schoener and Gorman, 1968; Hespenheide, 1971; Werner, 1974; Wilson, 1975; Zaret, 1980; Peters, 1983; Peters and Wassenberg, 1983; Gittleman, 1985; duToit and Owen Smith, 1989). Trophic level is not so important. Yes, tigers are rare and endangered, but so are many large herbivores like rhinos and tapirs.

If extreme guild members tend to be rare, perhaps we can better understand the extinction bias seen among mammal species in the late Pleistocene. In Australia, for example, most of the mammals that disappeared about 40 000 years ago were the largest of a group of near relatives (Murray, 1991; Flannery, 1994). Every species (about 40) weighing more than 60 kg disappeared. Some suspect that human hunting was the precipitating agent of this selective and far-reaching wave of extinctions (Martin and Klein, 1984). Quite possibly, but humans also hunt medium-sized species and most of them survived. It is true that modern extinctions in Australia have landed most heavily on medium-sized mammals (Burbidge and MacKenzie, 1989; Short and Smith, 1994), but this may be due to these extinctions coming in the wake of the megafaunal extinctions (little of the megafauna was left behind to destroy), and to the primary agents of modern mammal extinction being other medium-sized species (i.e. red foxes, feral cats and European rabbits).

5.2.4 Rarity and climate change – the compressed bits

The position of a species in its community often does depend upon a restricted set of habitats. That applies equally to species whose coexistence stems from distinct habitat preferences and to species whose coexistence stems from habitat tolerance variation in the face of shared preferences. Sometimes climate changes treat the world like an accordionist flexing his or her instrument, compressing the distributions of some habitats while expanding those of others. Species that depend on the compressed habitats get rare.

Webb (1984) notes the case of large North American mammals following the climate changes of the Late Miocene epoch. North America dried out, and its grasslands expanded at the expense of its savannas. Grazers proliferated. Twenty whole genera of browsers disappeared in the first wave of extinctions some 9 million years ago. But the climate change preceded the largest wave of extinctions by several million years. Between 5.4 and 6 million years ago (during the Pliocene epoch), 24 genera suddenly vanished. The species of those browsing genera must have been quite rare for a considerable length of time before their actual extinctions.

In the time since the Pliocene, not all species of grazers have done so well. *Mammuthus primigenius*, the woolly mammoth, grazed the tundra-steppes of Eurasia from Western Europe to the Pacific during the Pleistocene (Vartanyan *et al.*, 1993). By about 12 000 years ago, climatic change virtually eliminated those steppes south of 70°. The mammoth, its belly full of high Arctic grasses, suffered the same great restriction in distribution. By 9500 years ago, all relics of the once plentiful biome had disappeared from the mainland. So had the mammoth.

But the mammoth survived until only 3700 years ago in the one place the tundra-steppe did: Wrangel Island, Russia (Vartanyan *et al.*, 1993). Land to the south of Wrangel is too warm and too wet for this biome. In fact, by 7000 years ago mammoths had even evolved a dwarf form on Wrangel Island, so they must have done well enough to reproduce for many generations. But they must also have been quite rare. After all, a dwarf elephant is still an elephant. There could not have been all that many mammoths on Wrangel Island. Perhaps some plant species restricted to the tundra-steppe survive to this day as rarities limited to Wrangel Island.

Australia offers numerous examples of species whose ranges and abundances have expanded and contracted during the Pleistocene's pluvials and interpluvials. Keast (1961, 1981) has documented numerous birds now living in isolated pockets of moist coastal vegetation, but which, in cooler, wetter times, spread more expansively in the continent. The rarity of several birds is extreme. The noisy scrub-bird (Smith, 1985) provides a spectacular example. It now lives in a tiny corner of Western Australia where uncut, heathy and rarely burned forest–swamp ecotone survived both the end of the last glaciation and the attention of human managers (who are convinced that the only good forest is one set on fire every two or three years). Despite its apparency to census takers – conferred on it by its loud voice – the noisy scrub-bird is so rare (fewer than 200 pairs today; 25 years ago there were fewer than 50) that people thought it extinct for most of a century (Webster, 1962).

Using the fossil record, Coope (1979, 1987) traces the distribution of beetles through the advance and retreat of the ice in Eurasia. For example, both *Helophorus aspericollis* and *H. brevipalpis* have maintained their habitat preferences for at least tens of thousands of years. But when the ice advanced, so did *H. aspericollis*, the species that prefers cooler habitats.

When the ice retreated, leaving a more temperate climate, *H. brevipalpis* spread out. Species that are now rare and local – such as the dung beetle, *Aphodius holderi*, now confined to elevations above 3000 m in the Tibetan Plateau – were once spread over vast areas of Eurasia when those areas were colder. *Aphodius holderi* was the most common large dung beetle in Britain during the middle of the last glaciation.

Is there a way to predict from theories of community ecology which species are more likely to face habitat compression? We know of none. So this ecological problem may indeed affect species randomly.

5.2.5 Rarity and congenitally narrow habitat selection – the picky bits

Natural selection can reinforce narrow habitat selection (Rosenzweig, 1987a; Brown and Pavlovic, 1992), producing whole taxa of rare species with restricted distributions. Vrba (1987) has studied the fossil records of large mammals and discovered that some taxa do tend to have many species with narrow habitat requirements. In at least one case – the antelope subfamily Alcelaphini – this situation seems to have led to a 5-million-year record of rapid turnover of species, suggesting that species of Alcelaphini have tended to be rarer than most antelopes. The antelope genera *Damaliscus* (blesbuck, etc.) and *Connechaetus* (wildebeests) may also fit this pattern.

Some taxa composed of picky species have such specialized habitat requirements that natural selection makes them good inter-patch travellers. They must and do move from patch to patch in search of their special environment. Vrba notes the case of the aardvark whose need for sandy soil requires it to be able to find its habitat over a broad geographical range. Thus, what might have become several small ranges of rare species remains a single large range of a relatively common one. To produce rarity, congenital habitat pickiness must also restrict geographical range.

Even if geographical range is restricted so that many species evolve, a taxon may not exhibit the high turnover that suggests its species are rare. Both waterbucks and kudus sustain high diversity without having high turnover (see the re-analysis of Vrba's data in Rosenzweig and Vetault, 1992). Thus, this process remains speculative and poorly understood.

5.2.6 Rarity and centrifugal speciation – the marooned bits

The most accepted scenario of allopatric speciation maintains that small isolates undergo rapid evolutionary changes and tend to be the raw material of evolutionary novelty. They are thought to do so because of higher rates of stochastic deviation (drift, founder effect) and because of freedom from the constraint of the gene flow in their broadly dispersed ancestor. Given this scenario, new species may return to the areas of their ancestor and expand greatly. Such speciation would provide no clue to rarity.

But there is an alternative scenario that – were it true – would help. Brown (1957) theorized that peripheral isolates are conservative. They fail to adapt while, at the same time, their mother species moves along briskly, meeting the evolutionary challenges of a changing environment. Paradoxically, the isolates become new species because they do not change. Brown termed his theory 'centrifugal speciation'.

Some convincing evidence for centrifugal speciation exists. Much of it comes from the laboratory of Robert Baker, his colleagues and his students. It depends on the chromosomes and breeding behaviour of mice in the genus *Peromyscus*. *P. maniculatus* is a common species with a huge range that covers most of North America. On the periphery of that range, it has produced several new species with very small ranges (Blair, 1950). Examples include *P. polionotus* and *P. melanotis*, but the karyotypes of these two species closely resemble the common ancestor of all three. The karyotype of the large-range species, *P. maniculatus*, has changed much (Greenbaum *et al.*, 1978). Moreover, *P. melanotis* exists as a set of isolates on mountaintops in southeastern Arizona, New Mexico and Mexico. These tiny isolates do not appear to have diverged from each other in 10 000 years (Bowers *et al.*, 1973). Their karyotypes and their alleles match. Those in southeastern Arizona can interbreed amongst themselves and with other populations of *P. melanotis* from three Mexican populations. They have changed very little while their large range parent has evolved a lot.

The argument between centrifugal speciation and rapid change in isolates centres on scale. Rapid change in isolates depends on variation already present in a population. Genes and chromosomes mostly change their frequencies. In contrast, centrifugal speciation says that novelty depends on the appearance of new genes and karyotypes, surely a slower process. Walsh (1995), using concepts of molecular evolution completely unknown until the last decade, has theoretical evidence that further suggests the inability of small populations to produce much novelty. The incipient novelty of a gene duplication turns out to be much more likely to be neutralized in a small population than to be capitalized upon by the evolution of a new gene. The reverse is true of large populations.

Frey (1993, 1994) is developing cladistic methods to determine whether variation within species tends to agree with the centrifugal model or some other pathway of geographical speciation. She has applied these to three species of rodents (*Microtus montanus*, *Tamias minimus* and *Zapus hudsonius*). *Tamias* variation matched the predictions of centrifugal speciation well, although *Zapus* variation did not and the *Microtus* results have no unambiguous interpretation. We see no reason why Frey's methods should not also be applied among species. Will rare species often turn out to be conservative? Is rarity often a side effect of being abandoned in an ecological backwater, cut off from the evolutionary novelties that your relatives are using to adapt to a broad array of ecological problems in a wide variety of habitats?

5.2.7 Rarity and chaotic interactive dynamics – the poorly behaved bits

Ecological theory tells us that some species may exhibit chaotic dynamics. This does not mean that they have unpredictable population sizes. It means instead that their dynamics are governed by a strange attractor whose nature can be perfectly determined but only after decades or even centuries of work. Chaos emerges from the dynamic interactions of the species with its competitors and exploiters.

Vandermeer (1982) proposed that chaotic species are rare species. Though Rogers (1984) showed that the connection between chaos and rarity is not absolutely necessary, it seems undeniable that some chaotic species will often be rare. They will spend most of their generations in obscurity, but then suddenly erupt to populations that are orders of magnitude higher than usual.

It takes a long time to be sure of chaos. Many ecologists have yet to be convinced of chaos in even a single case. Nevertheless, if it does exist, theory tells us that we should expect some chaotic species to enter extended periods of quasi-equilibrium and mislead us into thinking that they are always going to be rare. Theory therefore does strongly suggest that chaos may be a primary cause of rarity. The record of a chaotic species will likely resemble that of the Pacific sardine. Starting in about 1920, the Pacific sardine supported a whole industry in Monterey, California. An army of fishermen harvested the abundance. Dozens of factories processed and packaged it. John Steinbeck described the era and its characters in his novel *Cannery Row*. And then they were gone – first the fish and then the industry.

Initially, most ecologists attributed the rarity of the Pacific sardine to over-fishing. But that appears to be, at most, only a minor part of the story (Soutar and Isaacs, 1974). Instead, this fish exhibits dynamics that seem chaotic.

We learn about its population trajectories only from subfossil deposits of fish scales that generate estimates of population size (Figure 5.3; Soutar and Isaacs, 1969). Other fish populations wobble along throughout the centuries as if they were cycling or at steady state, but the sardine does not. Most decades it is rare. In over 55% of the decades from the year 160, it was so rare as to be virtually absent from core samples in the Santa Barbara Basin. It is even rarer in a similar basin off the coast of Baja California (Soutar and Isaacs, 1974). Every so often, however, it explodes. Soutar and Isaacs (1969) estimate minimum population levels during the largest peak of the last 1900 years at 30 000 sardines/km^2. Typical peaks have had about half that density. There seems to be no periodicity to these

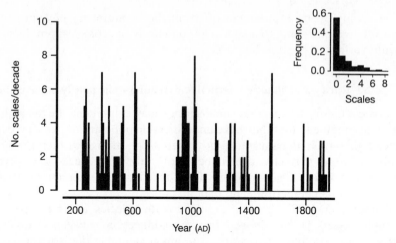

Figure 5.3 Population of the Pacific sardine for the past two millennia. Each datum comes from the density of fish scales in a marine petroleum exploration core during a 1850-year sequence of 10-year periods. Modern work with such scales shows that their densities correlate very strongly with fish population size. Other cores in the same and another basin produced similar fluctuations, peaking and declining simultaneously. (After Soutar and Isaacs, 1969.)

population explosions. Intervals between peaks range from 20 to 200 years. Peaks last for 20–150 years. Despite all the research on the sardine, no model and no data predicted that 'pattern'.

5.2.8 Rarity and recent introductions – the tragic bits

No matter how great its current abundance, no species is immune to rarity. Immigrant species can move suddenly into a biogeographical province and lay common species low. Extraordinarily common species have imploded from such a novelty, losing most or all of their abundance within a single human lifetime. Sometimes the new species appeared naturally. Sometimes it was transported by human agency. And sometimes we ourselves were the problem. We know too many examples.

The limpet, *Lottia alveus*, provides a natural example (Carlton *et al.*, 1991). Once it was common in eelgrass beds on the coast of the western Atlantic Ocean from Egg Harbor, Labrador to Long Island Sound, New York. It ate only eelgrass, *Zostera marina*, and lived only in fully marine salinities (not estuaries). Despite its high degree of specialization, it was abundant where found.

In 1930–1933, *Zostera marina* suddenly vanished from vast areas of the North Atlantic. It survived only in water of low salinity. The decimation of eelgrass – probably because of a new slime mould disease – had many severe effects. But *Lottia alveus* actually vanished forever; its niche, the combination of marine salinity and eelgrass, had been obliterated from its geographical range.

Do not conclude that *Lottia alveus* was doomed solely because of its narrow niche. The same fate overtook one of North America's most successful and adaptable tree species, the American chestnut (*Castanea dentata*). A century ago, chestnuts made up about 25% of all forest trees in their large range (Roane *et al.*, 1986). That range extended from southern Mississippi and Georgia to Maine, and from Delaware to southeastern Michigan and southern Illinois. They were abundant even in forests where they did not quite belong. For example, in a beech–maple forest near Cleveland, Ohio, chestnuts accounted for 14.9% of all trees wherever the land rolled (Williams, 1936).

While they are small trees, chestnuts have smooth, shiny bark that is green and full of chloroplasts. Rutter (personal communication) believes the chloroplasts indicate wintertime photosynthesis and perhaps even wintertime growth. So they are active in all sorts of times as well as places. In fact, during better years and in better places, rapid summer growth produces most of the wood. Summer wood has almost no vessels, so it is dense and hard. In poorer situations, spring-grown wood forms a higher proportion of the tree. Spring growth has the vascular system, so it is soft and light. Thus, the chestnut constitutes both the softwoods and hardwoods of its forest.

Older chestnuts have deeply furrowed, thick bark that resists fire, and they are the predator-resistant species *par excellence*. Their bark and wood have very high concentrations of tannins. Their nuts lie virtually impregnable in a strong, spiny pod. Only a species or two of co-evolved weevil can eat chestnuts in the pod. Even soil type seems not to matter. Their soil can range from clay to sand.

American chestnuts were also the most formidable competitors in their bailiwick (Paillet and Rutter, 1989). They grow towards – and through – a canopy of oak (*Quercus*) and big-tooth aspen (*Populus*) as if it were a light gap. Evidence suggests that they also make a pathogenic chemical with which to fight their competitors.

Nevertheless, they survive only in a few isolated stands and as root crowns of old chestnuts that repeatedly send new sprouts into the forest where they are doomed to sicken and die of chestnut blight, *Endothia parasitica* – the disease responsible for their decimation. Chestnut blight is a new disease probably introduced to North America by human commerce. It is a fungus that girdles and destroys its victims. It killed about 3.5 billion chestnuts in the 50 years after it was discovered by Merkel (1906).

There is no doubt that human beings caused the rarity and eventual extinction of one of the earth's most abundant vertebrate species, the passenger pigeon (*Ectopistes migratorius*). Although people had not harmed this species for 10 000 years, the development and introduction of western farming methods and hunting tools reduced the world's most abundant bird to rarity and then extinction within a single century. Schorger (1955) estimates the total population of passenger pigeons at three to five billion when the USA was young. That is much more likely to be an underestimate than an exaggeration (Rosenzweig, 1995). Today, all other species of breeding birds in the USA put together have about 5.7 billion individuals. Thus, two centuries ago, every other bird in the USA was a passenger pigeon.

At times, species get free of some of their negative interactions and become pests or weeds. Often the result is a large increase in abundance that shows us just how rare these interactions can make them. The literature of biological control contains many examples of the suppression of such species by the careful use of their enemies. A classic case comes from Australia. Two species of prickly pear cactus (*Opuntia inermis* and *O. stricta*), introduced without their predators, at first rampaged over economically important parts of the continent. Then a predator (a moth, *Cactoblastis cactorum*) was also introduced and brought them under control (Dodd, 1959).

The history of Monterey pine, *Pinus radiata*, also shows how negative interactions may make a dramatic difference to distributions and population sizes. In the temperate parts of the southern hemisphere *Pinus radiata* has huge economic value. Foresters grow it more than any other timber tree. It grows straight and thick and tall and fast. But in its native California it is rare and short – useless for industry. The difference is thought to be its natural consumers: they have not yet made the trip south. One can only wonder what the distribution and abundance of Monterey pine would be in the USA were some of the species' enemies to disappear. Yet there may not be much use in such speculation. Kennedy and Southwood (1984) and Rosenzweig (1995) looked at the species diversities of insects specialized on British trees. It takes only a few centuries for the full complement of a tree's 'associates' to follow their host to a new biogeographical site. After that time has passed, why do some trees then keep growing straight and tall and cover the landscape with individuals, while others become rare? Perhaps this too is merely accidental?

5.3 DISCUSSION

5.3.1 Habitat scarcity is often – but not always – implicated in rarity

We have concentrated on setting out the ecological processes that may make species rare. No one will be surprised that most tell how a species' habitat becomes rare. A species can inherit – as a taxonomic characteristic – the tendency to depend on a narrow set of habitats. Or it can evolve that tendency itself by becoming the master of a rich but rare habitat. Or – without evolution – its habitat can lose ground in a climate change. It can even lose some of its habitat to an invading competitor.

Rarity does not always mean habitat scarcity. The invader may be a predator, for example. Whereas in the case of eelgrass the predator also restricted its victim's habitat (eelgrass lost access to water of marine salinity), many predators will follow their victim into all its habitats (as *Endothia* does to American chestnut).

Even without predation, habitat may not figure in the reckoning. For example, the niches of some species require specialization on a relatively uncommon resource. Such species could be widespread, but sparse (as was the American burying beetle). Species governed by chaotic dynamics may or may not be narrowly distributed. They may well be narrowly distributed when they are rare, but we suspect they will not be during outbreaks; otherwise they would have to violate Fretwell's principle that abundant species must spread out into many habitat types.

Fretwell's principle hints at a caveat for explorers of rarity: a species that is rare often needs to restrict itself to a small number of habitats (Fretwell and Lucas, 1970). Thus, if we simply survey and find that rare species use few habitats, we will not know much. Maybe, if they were more abundant, they would use more! Is this partly the reason for Hanski's pattern (Hanski *et al.*, 1993)? Are rare species less widespread geographically because they concentrate in only a few habitats? Or does Hanski's pattern operate at a much larger scale than Fretwell's? We leave that puzzle for the future, but emphasize it because we are unaware that anyone has previously mentioned it.

Meanwhile, what about a product of centrifugal speciation? Is such a relict rare because it lacks enough habitat? We suspect the answer to be more semantic than biological. Surely, it is a relict because its habitat occurred in a small patch too isolated for panmixia with its mother species. So, local habitat rarity is implicated, but its habitat probably remained quite common at a larger scale (otherwise its mother would also be rare). Thus, its real problem is access.

Finally, should it turn out that the hypothesis 'rarity accompanies high trophic level' is correct after all, we will have yet another case of rarity without habitat restriction.

5.3.2 The importance of intolerance

Habitat intolerance is easy to document. Moreover, it is related to other notions of community organization that themselves appear well established. Many cases of experimental investigation into competition show asymmetry, for instance (Lawton and Hassell, 1981). In asymmetry, one species has a strong competitive effect on the other, but the reverse effect is weak or absent. That is a hallmark of shared preference or tolerance organization. The intolerant has the strong effect on the tolerant.

Rosenzweig (1987b: 482) traced shared preferences to quantitative niche axes. A quantitative niche axis is 'an array of habitats whose resources are the same and are obtained with the same foraging techniques. The habitats vary . . . only in the amounts of resources they offer.' That is why all the species prefer the same habitat.

The concept of the quantitative axis has also been used by students of plant–animal interactions. Both Rhoades and Cates (1976) and Feeny (1976) suggested that some plant species vary in their defences by having different concentrations of nasty chemicals, rather than protecting themselves with a variety of truly lethal chemicals. Naturally, all consumers can be expected to prefer the plant tissues with the small concentrations; hence, this too should yield a shared preference guild. But we do not know whether the intolerant species of such guilds also find themselves rare. Presumably, they should if plants and plant tissues with low concentrations of nasty chemicals find themselves scarce because they attract too much consumer attention. This is not improbable at all.

Grime's (1977) triangle for plant specializations also contains a well established quantitative axis. Grime calls it the stress axis, but defines it clearly as a productivity axis. In his scheme, species adapted to the rich end of the stress axis have to be good competitors, but they require both high productivity and a low disturbance rate. Does that also tend to make them rare? It would not seem unlikely.

Tolerance variation may also explain some mysterious results in the literature of competition; for example, those of Lawton (1984) on fern-eating insects. Lawton based his work on a distinct-preference theory of competitive coexistence. This theory is the one generally seen in basic ecology textbooks. In it, each species has a specialized niche different from any other. If the specialty is a habitat specialty, then few habitats will support both species for long.

Lawton used that prediction to set up working hypotheses for the joint distribution of fern-eating insects. He designated matched pairs of species in which both species attack the fern similarly and thus appear to be probable competitors. Often, matched pairs include two species of the same genus. Lawton predicted that patches of bracken would tend to harbour one or the other species of a matched pair, but not both.

Lawton's working hypothesis therefore was that the two species of a matched pair should rarely be found together. He tested this hypothesis with a set of 2 × 2 contingency tables. In many cases, the table deviated from randomness, but the deviations did not follow the working hypothesis.

Figure 5.4 shows four of the non-random tables. In each you will see that the cell with both species is well represented. The source of the significant departure from randomness is the small number of times that one of the species occurs alone. While this result does not agree with distinct-preference coexistence, it agrees perfectly with coexistence based on variation in tolerance. The under-represented cells should be those which have only the intolerant species. This explanation can be tested experimentally.

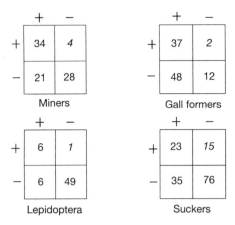

Figure 5.4 Numbers of locations with different combinations of bracken-fern-eating insect species pairs. The contingency tables indicate presence (+) and absence (−) of a species – one species by the rows, the other by the columns. In each table we entered the significant number – the one significantly less than the null hypothesis – in italics in the upper righthand cell. It is always one of the two with only one species. As Lawton points out, this result contradicts the assumption that potential competitor species specialize on different habitats, but it does agree with the idea that these species coexist as competitors by having tolerance differences. (Data from Lawton, 1984.)

The tilt that intolerant species give to abundance and geographical range may underlie the correlation between them, and supply the mechanism for Brown's (1984) hypothesis. That hypothesis points to niche breadth as the explanation for the Hanski pattern. Previous attempts to validate it (Hanski *et al.*, 1993; Lawton, 1993) may have failed because they looked too broadly at the data base and did not concentrate on guilds of species organized by tolerance differences. The habitats of tolerant species should be more common and widespread both because tolerant species use more habitats and because these habitats occupy the mid-range of a frequency distribution. If there is no trend among exceptions to the rule, any correlation will come from whatever signal – however weak – emanates from those that follow the rule.

5.3.3 Is there an effect of trophic level *per se*?

One of our surprises in preparing this chapter was the result we obtained from analysing the trophic level hypothesis. When Kunin first solicited our contribution for this book, he hoped we would not restrict ourselves to such old and well accepted ideas as 'rarity accompanies high trophic level'. We agreed that we wished to explore new ground, but felt obliged to document the trophic level pattern. As you now know, this pattern is not so easy to document, perhaps because it does not exist. Recognizing that higher trophic levels have fewer species resulted in our estimating that trophic level may play no role in setting the amount of energy that flows to a single species. If so, trophic level would play no role in determining abundances. How have the mighty fallen!

5.3.4 Some possible future additions to the list

(a) Certain insect herbivores

Two other sources of rarity – chaos and congenitally picky habitat selection – have possible parallels in an unexploited paper by Rhoades (1985). He suggests that insect herbivores may be divided into opportunists and stealth specialists. By definition, opportunistic herbivores specialize on well defended plant species. When, unusually, the plant's defences are weak, the opportunist population explodes. Thus, opportunists have highly variable population sizes. Is there a causal connection between such a pattern and chaos? We do not know.

Rhoades's alternative herbivore strategy, stealth, leads to consumers with small but steady population sizes. Stealthy consumers suffer the defences of plants at full strength but have the ability to withstand them, or even to capitalize on them. Stealthy consumers stay relatively rare because they keep their victims rare. Do stealthy consumers parallel congenitally picky habitat selectors? Is there something that predisposes the members of some consumer taxa to specialize on being able to eat a poison or two with impunity – or even with relish? We should point out that Rhoades's categories need to be confirmed and explored.

Opportunists and stealth specialists would add to the list of ways that exploitative interactions can produce rarity. Even so, they will probably not complete the list. Ecologists may be able to discover a systematic way to predict which victim species are most likely to be held to rarity after they have reached an evolutionary steady-state with their enemies.

(b) Body size

The influence of body size, *per se*, on abundance constitutes another category that might be explored for additional causes of rarity. We have touched briefly on the issue of body size in several places in the chapter. Large body size often allows a species to dominate the thick bits – those that represent rich, rare habitats. In some taxa, large size may correlate with high trophic level – the high bits. In others, extreme size may be associated with extreme guild position – the end bits.

Ecologists have not reached consensus about the overall relationship of body size to abundance, probably because it is a complex and difficult subject. In many taxa, population density declines with increase in body size (e.g. Damuth, 1981, 1987; Peters and Raelson, 1984; Lawton, 1989;

Blackburn *et al.*, 1993). The relationship may be weak and explain little of the variance, but that is because small species have all sorts of densities; big species are always sparse (Lawton, 1989). Nevertheless, abundance is the product of density times geographical range. Do big species have larger geographical ranges that offset their low densities? We remain unsure about the overall relationship of size to abundance.

Even if we knew the pattern of body size *vs.* abundance, we would remain ignorant about its ecological causes (that is, if there is pattern and it does have such causes). So, except for the three points mentioned at the start of this subsection, we consciously did not attempt to treat the issue of body size here. Once the pattern becomes clearer, we should be able to discover the ecological mechanisms that connect body size to rarity.

5.3.5 Is rarity predictable?

Two of our causes of rarity – intolerance and congenital habitat pickiness – seem more predictable than the others. Rarity from congenital habitat pickiness is predictable by definition. Once a taxon acquires the inherited tendency toward habitat pickiness, its constituent descendants are likely to keep it. Picky taxa have high turnover rates and one day we may learn that those rates are – at least partly – under the control of the species' ecologies.

Rarity from intolerance is predictable for a different reason (Rosenzweig, 1989). Evolution toward intolerance is a case of conflict between the good of the group and that of the individual. Intolerant species need and take the richest places, i.e. the best spots for individuals. But if that makes the species rare, it will increase its chance of extinction. So natural selection may push tolerant species toward intolerance, rarity and a higher extinction rate. Rosenzweig (1989) compared evolution in shared preference guilds to a parade. Intolerant descendants disappear and tolerant ones are drawn into their niche by the selfishness of natural selection. Some day, we ought to be able to point to a shared-preference species and foretell its doom. Moreover, we probably will be able to estimate how long that fate will take to arrive and which species in the guild will supplant it as the intolerant. We should be able to see this in the fossil record as a taxon cycle. Rarity in shared–preference guilds seems almost as predictable as nightfall.

A third cause of rarity may also be predictable, though at a community level. Guilds whose species specialize by having different body sizes will find themselves generating some extreme-sized species – species that tend toward rarity. We can imagine that, once it has achieved a better understanding of the coevolution of niches, ecology will be able to predict how many.

The other four sources of rarity seem mostly unpredictable. Maybe we can say that a common, widespread species is likely to have some relictual, rare and isolated daughters. But how many? And when? The answers would seem to depend on geology and other processes extraneous to community ecology – and predicting the others seems like a pipe-dream. Which species will lose habitat to an ice age? Which will be governed by chaos? Which will fall to an invader? We do not know if we should ever expect ecology to have the power to answer any of these questions.

5.3.6 Consequences for conservation

In large part, we want to know about rarity to help to slow extinction down. Is it so important to conservation efforts that many of the processes we have discussed may not be predictable? Perhaps being able merely to identify the process that has produced a species' rarity can help us to protect it. Here we have more hope.

(a) Protecting species falling to an invader

We can adopt one or two tactics. We can try to increase the rate at which it develops resistance to the invading enemy. We can also try to control the invader with another species or interfere with its life cycle. Such human interference is often mustered to control plant and insect pests, but it could also be used for conservation.

(b) Protecting intolerant species

If a species is naturally rare because it is intolerant, we can also use pest-control tools, but on its competitor. This will be much more difficult than controlling a weed. First, we must learn the identity of the tolerant(s). Almost certainly, they will be rather closely related to the intolerant rare species. Then we must show by experiment that the rare species retains the behavioural flexibility to expand its use of habitats should its tolerant competitors get scarcer (occasionally, a natural experiment – the expansion of the intolerant's habitat niche when or where the tolerant is missing or reduced in population – will serve to substitute for an experiment). If the rare species does retain the flexibility to expand its use of habitats, then we must find a way to reduce the tolerant's population size without decimating it. Yes, we will need to treat the tolerant as if it were a pest. Worse, it may look and act much like the intolerant species, making difficult the job of finding an agent that can distinguish between them. Worse still, we will not want to eliminate or even endanger the tolerant, so we will have to find a scalpel rather than a butcher's knife. It may be quite difficult to find a control agent specific enough to inhibit the tolerant gently without harming the intolerant too. Nevertheless, our goal will be clear and our path well marked.

(c) Protecting chaotically rare species

The work of Allen *et al.* (1993) suggests that chaos might actually help to prevent extinction in metapopulations. Even if that is not always true, chaos may engender adaptations that coincidentally protect chronically rare species from extinction. So, maybe a chaotically rare species does not need our help. However, the same argument may apply to any naturally chronically rare species. Look at the information about chronically rare species in Chapters 11 and 14. Does it suggest that species that have had a chance to adapt to chronic rarity should tend to need less conservation effort? We do not believe so.

(d) Protecting other kinds of rare species

Maybe a centrifugally produced relictual species does not warrant our help – it is an evolutionary backwater anyway. Maybe a species whose habitat has shrunk because of climate change is beyond help and all we can do is wave it a kiss goodbye. Perhaps the same fate must inevitably overtake a species whose niche lies at the fringe of its guild's adaptive zone when that guild's overall abundance begins to shrink from competition with humanity. And perhaps there is nothing we can do for a congenitally narrow habitat selector whose guild is contracting in diversity because of habitat loss? We do not yet know. But we can guess that conservation efforts and resources may need to be directed in the most efficient channels. Any responsible strategy for conservation will have to resolve such questions before diverting those resources to such species.

We already know that we ought to be able to extend special help to intolerant rarities and to species suffering because of some invader that we carted in from a foreign biogeographical province. And when the stakes are very high – as often they are in the case of a spectacular and charismatic species – we can point out in advance if a case is unlikely to have a satisfactory outcome, thus minimizing the disillusionment that is bound to accompany our failure to become omnipotent.

Let us reiterate the question that started this chapter. Is there any use trying to understand what makes species become rare and go extinct? Or do species evanesce like molecules of hot water in a coffee cup? Our exploration of the question leads us to believe that seeking the ecological causes of rarity does have value. We are somewhat encouraged that, even if predictability is only a partly attainable goal, the causes of rarity may be relatively few and perhaps not too hard to identify. And in two cases – both of which we believe to be important – the processes suggest practical methods of amelioration that do not involve buying up the whole earth for a National Park.

ACKNOWLEDGEMENTS

Thanks to Bill Kunin and John Lawton for their helpful comments and questions. NSF grant BSR-8905728 helped to support the work of M.L.R.

REFERENCES

Abramsky, Z., Rosenzweig, M.L., Pinshow, B. *et al.* (1990) Habitat selection: An experimental field test with two gerbil species. *Ecology*, **71**, 2358–2369.

Allen, J.C., Schaffer, W.M. and Rosko, D. (1993) Chaos reduces species extinction by amplifying local population noise. *Nature*, **364**, 229–232.

Anderson, R.S. (1982) On the decreasing abundance of *Nicrophorus americanus* Olivier (Coleoptera: Silphidae) in eastern North America. *Coleopterists Bulletin*, **36**, 362–365.

Ashmole, N.P. (1968) Body size, prey size, and ecological segregation in five sympatric tropical terns (Aves: Laridae). *Systematic Zoology*, **17**, 292–304.

Blackburn, T.M., Brown, V.K., Doube, B.M. *et al.* (1993) The relationship between abundance and body size in natural animal assemblages. *Journal of Animal Ecology*, **62**, 519–528.

Blair, W.F. (1950) Ecological factors in the speciation of *Peromyscus*. *Evolution*, **4**, 253–275.

Bovbjerg, R.V. (1970) Ecological isolation and competitive exclusion in two crayfish (*Orconectes virilis* and *Orconectes immunis*). *Ecology*, **51**, 225–236.

Bowers, J.H., Baker, R.J., and Smith, M.H. (1973) Chromosomal, electrophoretic and breeding studies of selected populations of deer mice (*Peromyscus maniculatus*) and black-eared mice (*P. melanotis*). *Evolution*, **27**, 378–386.

Brown, J.H. (1984) On the relationship between abundance and distribution of species. *American Naturalist*, **124**, 225–279.

Brown, J.H., Marquet, P.A. and Taper, M.L. (1993) Evolution of body size: consequences of an energetic definition of fitness. *American Naturalist*, **142**, 573–584.

Brown, J.S. (1989) Desert rodent community structure: a test of four mechanisms of coexistence. *Ecological Monographs*, **59**, 1–20.

Brown, J.S. and Pavlovic, N.B. (1992) Evolution in heterogeneous environments: effects of migration on habitat specialization. *Evolutionary Ecology*, **6**, 360–382.

Brown, W.L. Jr. (1957) Centrifugal speciation. *Quarterly Review of Biology*, **32**, 247–277.

Burbidge, A.A. and McKenzie, N.L. (1989) Patterns in the modern decline of Western Australia's vertebrate fauna: causes and conservation implications. *Biological Conservation*, **50**, 143–198.

Carlton, J.T., Vermeij, G.J., Lindberg, D.R. *et al.* (1991) The first historical extinction of a marine invertebrate in an ocean basin: the demise of the eelgrass limpet *Lottia alveus*. *Biological Bulletin*, **180**, 72–80.

Chase, J.M. and Belovsky, G.E. (1993) Experimental evidence for the included niche. *American Naturalist*, **143**, 514–527.

Cohen, J.E., Pimm, S.L., Yodzis, P. and Saldaña, J. (1993) Body sizes of animal predators and animal prey in food webs. *Journal of Animal Ecology*, **62**, 67–78.

Colwell, R.K. and Fuentes, E.R. (1975) Experimental studies of the niche. *Annual Review of Ecology and Systematics*, **6**, 281–310.

Connell, J.H. (1961) The influence of interspecific competition and other factors on the distribution of the barnacle *Chthamalus stellatus*. *Ecology*, **42**, 710–723.

Coope, G.R. (1979) Late cenozoic fossil Coleoptera: Evolution, biogeography, and ecology. *Annual Review of Ecology and Systematics*, **10**, 247–267.

Coope, G.R. (1987) The response of late Quaternary insect communities to sudden climatic changes, in *Organization of Communities Past and Present* (eds J.H.R. Gee and P.S. Giller), Blackwell Scientific, Oxford, pp. 421–438.

Crawley, M.J. (ed.) (1992) *Natural Enemies*, Blackwell Scientific Publications, Oxford, UK.

Damuth, J. (1981) Population density and body size in mammals. *Nature*, **290**, 699–700.

Damuth, J. (1987) Interspecific allometry of population densities in mammals and other animals: the independence of body mass and population energy use. *Biological Journal of the Linnaean Society*, **31**, 193–246.

Davidson, D.W. (1977) Species diversity and community organization in desert seed-eating ants. *Ecology*, **58**, 711–724.

Dodd, A.P. (1959) The biological control of prickly pear in Australia, in *Biogeography and Ecology in Australia. Monographiae Biologicae*, **8** (eds A. Keast, R.L. Crocker and C.S. Christian), Dr W. Junk, The Hague, pp. 565–577.

duToit, J.T. and Owen-Smith, N. (1989) Body size, population metabolism, and habitat specialization among large African herbivores. *American Naturalist*, **133**, 736–740.

Feeny, P. (1976) Plant apparency and chemical defense, in *Biochemical interactions between plants and insects* (eds. J.W. Wallace and R.L. Mansell), Plenum, New York pp 1–40.

Flannery, T.F. (1994) *The Future Eaters; an ecological history of the Australian land and people*, Reed Books, Chatswood, NSW.

Fretwell, S.D. and Lucas, H. Jr (1970) On territorial behavior and other factors influencing habitat distribution in birds. *Acta Biotheoretica*, **14**, 16–36.

Frey, J.K. (1993) Modes of peripheral isolate formation and speciation. *Systematic Biology*, **42**, 373–381.

Frey, J.K. (1994) Testing among modes of allopatric speciation: a hypothetico-deductive approach. PhD dissertation, University of New Mexico.

Gittleman, J.L. (1985) Carnivore body size: ecological and taxonomic correlates. *Oecologia*, **67**, 540–544.

Greenbaum, I.F., Baker, R.J. and Ramsey, P.R. (1978) Chromosomal evolution and the mode of speciation in three species of *Peromyscus*. *Evolution*, **32**, 646–654.

Grime, J.P. (1977) Evidence for the existence of three primary strategies in plants and its relevance to ecological and evolutionary theory. *American Naturalist*, **111**, 1169–1194.

Hanski, I. (1982) Dynamics of regional distribution: the core and satellite species hypothesis. *Oikos*, **38**, 210–221.

Hanski, I., Kouki, J. and Halkka, A. (1993) Three explanations of the positive relationship between distribution and abundance of species, in *Species Diversity in Ecological Communities: historical and geographical perspectives*, (eds R. Ricklefs and D. Schluter), University of Chicago Press, Chicago, pp. 108–116.

Hespenheide, H.A. (1971) Food preference and the extent of overlap in some insectivorous birds, with special reference to Tyrannidae. *Ibis*, **1134**, 59–72.

Holmes, J.C. (1961) Effects of concurrent infections on *Hymenolopsis diminuata* (Cestoda) and *Monoliformis dubius* (Acanthocephala). I. General effects and comparison with crowding. *Journal of Parasitology*, **47**, 209–216.

Keast, A. (1961) Bird speciation on the Australian continent. *Bulletin of the Museum of Comparative Zoology*, **123**, 303–495.

Keast, A. (1981) The evolutionary biogeography of Australian birds, in *Ecological Biogeography of Australia. Monographiae Biologicae*, **41**, (ed. A. Keast), Dr W. Junk, The Hague, pp.1585–1635.

Kennedy, C.E.J. and Southwood, T.R.E. (1984) The number of species of insects associated with British trees: a re-analysis. *Journal of Animal Ecology*, **53**, 455–478.

Kozol, A.J., Scott, M.P. and Traniello, J.F.A. (1988) The American burying beetle, *Nicrophorus americanus*: studies on the natural history of a declining species. *Psyche*, **95**, 167–176.

Lawton J.H. (1984) Non-competitive populations, non-convergent communities, and vacant niches: The herbivores of bracken, in *Ecological Communities: Conceptual Issues and the Evidence*, (eds D.R. Strong, D.S. Simberloff, L.G. Abele and A.B. Thistle), Princeton University Press, Princeton, NJ, pp. 67–100.

Lawton, J.H. (1989) What is the relationship between population density and body size in animals? *Oikos*, **55**, 429–434.

Lawton, J.H. (1993) Range, population abundance and conservation. *Trends in Ecology and Evolution*, **8**, 409–413.

Lawton, J.H. and Hassell, M.P. (1981) Asymmetrical competition in insects. *Nature*, **289**, 793–795.

Lomolino, M.V., Creighton, J.C., Schnell, G.D. and Certain, D.L. (1995) Ecology and conservation of the endangered American burying beetle (*Nicrophorus americanus*). *Conservation Biology*, **9**, 605–614.

Martin, P.S. and Klein, R.G. (eds), (1984) *Quaternary Extinctions: a prehistoric revolution*, University of Arizona Press, Tucson.

May, R.M. (1975) Patterns of species abundance and diversity, in *Ecology and Evolution of Communities*, (eds M.L. Cody and J.M. Diamond), Belknap Press of Harvard University Press, Cambridge, Mass., pp. 81–120.

McGowan, J.A. and Walker, P.W. (1985) Dominance and diversity maintenance in an oceanic ecosystem. *Ecological Monographs*, **55**, 103–118.

McGowan, J.A. and Walker, P.W. (1993) Pelagic diversity patterns, in *Species Diversity in Ecological Communities: historical and geographical perspectives*, (eds R. Ricklefs and D. Schluter), University of Chicago Press, Chicago, pp. 203–214.

Merkel, H.W. (1906) A deadly fungus on the American chestnut. *NY Zoological Society Annual Report*, **10**, 97–103.

Murray, P. (1991) The Pleistocene megafauna of Australia, in *Vertebrate Palaeontology of Australasia*, (eds P. Vickers-Rich, J.M. Monaghan, R.F. Baird *et al.*), Pioneer Design Studio and Monash University Publications Committee, Melbourne, pp. 1071–1164.

Paillet, F.L. and Rutter, P.A. (1989) Replacement of native oak and hickory tree species by the introduced American chestnut (*Castanea dentata*) in southwestern Wisconsin. *Canadian Journal of Botany*, **67**, 3457–3469.

Pearson, D.L. and Mury, E.J. (1979) Character divergence and convergence among tiger beetles (Coleoptera: Cicindelidae). *Ecology*, **60**, 557–566.

Peters, R.H. (1983) *The Ecological Implications of Body Size*, Cambridge University Press, Cambridge, England.

Peters, R.H. and Raelson, J.V. (1984) Relations between individual size and mammalian population density. *American Naturalist*, **124**, 498–517.

Peters, R.H. and Wassenberg, K. (1983) The effect of body size on animal abundance. *Oecologia*, **60**, 89–96.

Pimm, S.L., Rosenzweig, M.L. and Mitchell, W. (1985) Competition and food selection: field tests of a theory. *Ecology*, **66**, 798–807.

Rabinowitz, D. (1981) Seven forms of rarity, in *The Biological Aspects of Rare Plant Conservation*, (ed. H. Synge), John Wiley and Sons, NY, pp. 205–217.

Rhoades, D.F. (1985) Offensive–defensive interactions between herbivores and plants: their relevance in herbivore population dynamics and ecological theory. *American Naturalist*, **125**, 205–238.

Rhoades, D.F. and Cates, R.G. (1976) Towards a general theory of plant anti-herbivore chemistry, in *Biochemical Interactions Between Plants and Insects* (eds J.W. Wallace and R.L. Mansell), Plenum, New York, pp.168–213.

Roane, M.K., Griffin, G.J. and Elkins, J.R. (1986) *Chestnut Blight, Other Endothia Diseases, and the Genus Endothia*. American Phytopathological Society, St Paul, MN.

Roff, D. (1981) On being the right size. *American Naturalist*, **118**, 405–422.

Rogers, T.D. (1984) Rarity and chaos. *Mathematical Biosciences*, **72**, 13–17.

Rosenzweig, M.L. (1966) Community structure in sympatric Carnivora. *Journal of Mammalogy*, **47**, 602–612.

Rosenzweig, M.L. (1974) On the evolution of habitat selection. *Proceedings of the First International Congress of Ecology*, pp. 401–404.

Rosenzweig, M.L. (1979) Optimal habitat selection in two-species competitive systems, in *Population Ecology; Fortschritt in Zoologie*, **25**, (eds U. Halbach, and J. Jacobs), Gustav Fischer, Stuttgart, pp. 283–293.

Rosenzweig, M.L. (1987a) Habitat selection as a source of biological diversity. *Evolutionary Ecology*, **1**, 315–330.

Rosenzweig, M.L. (1987b) Community organization from the point of view of habitat selectors, in *Organization of Communities Past and Present*, (eds J.H.R. Gee and P.S. Giller), Blackwell Scientific, Oxford, pp. 469–490.

Rosenzweig, M.L. (1989) Habitat selection, community organization, and small mammal studies, in *Patterns in the Structure of Mammalian Communities*, (eds D.W. Morris, B.J. Fox and Z. Abramsky), Texas Tech University Press, Lubbock, TX, pp. 5–21.

Rosenzweig, M.L. (1991) Habitat selection and population interactions: the search for mechanism. *American Naturalist*, **137**, S5–S5.

Rosenzweig, M.L. (1995) *Species Diversity in Space and Time*, Cambridge University Press, Cambridge, England.

Rosenzweig, M.L. and Vetault, S. (1992) Calculating speciation and extinction rates in fossil clades. *Evolutionary Ecology*, **6**, 90–93.

Schoener, T.W. and Gorman, G.C. (1968) Some niche differences in three Lesser Antillean lizards of the genus *Anolis*. *Ecology*, **49**, 819–830.

Schoenly, K., Beaver, R.A. and Heumier, T.A. (1991) On the trophic relations of insects: a food-web approach. *American Naturalist*, **137**, 597–638.

Schorger, A.W. (1955) *The Passenger Pigeon: its natural history and extinction*, University of Wisconsin, Madison.

Short, J. and Smith, A. (1994) Mammal decline and recovery in Australia. *Journal of Mammalogy*, **75**, 288–297.

Smith, G.T. (1985) The noisy scrub-bird *Atrichornis clamosus*. Does its past suggest a future?, in *Birds of Eucalypt Forests and Woodlands: Ecology, conservation, management*, (eds A. Keast, H.F. Recher, H. Ford and D. Saunders), Royal Australasian Ornithologists Union and Surrey Beatty and Sons, Chipping Norton, NSW, pp. 301–308.

Soutar, A. and Isaacs, J.D. (1969) History of fish populations inferred from fish scales in anaerobic sediments off California. *California Cooperative Oceanic Fisheries Investigations Reports*, **13**, 63–70.

Soutar, A. and Isaacs, J.D. (1974) Abundance of pelagic fish during the 19th and 20th centuries as recorded in anaerobic sediment off California. *Fishery Bulletin*. **72**, 257–274.

Stanley, S.M. (1979) *Macroevolution. Pattern and process*, W.H. Freeman, San Francisco.

Storer, R.W. (1966) Sexual dimorphism and food habits in three North American accipiters. *Auk*, **83**, 423–436.

Vandermeer, J. (1982) To be rare is to be chaotic. *Ecology*, **63**, 1167–1168.

Vartanyan, S.L., Garutt, V.E. and Sher, A.V. (1993) Holocene dwarf mammoths from Wrangel Island in the Siberian Arctic. *Nature*, **362**, 337–340.

Vézina, A.F. (1985) Empirical relationships between predator and prey size among terrestrial vertebrate predators. *Oecologia*, **67**, 555–565.

Vrba, E.S. (1987) Ecology in relation to speciation rates: some case histories of Miocene – Recent mammal clades. *Evolutionary Ecology*, **1**, 283–300.

Walsh, J.B. (1995) How often do duplicated genes evolve new functions? *Genetics*, **139**, 421–428.

Warren, P.H. and Lawton, J.H. (1987) Invertebrate predator–prey body size relationships: an explanation for upper triangular food webs and patterns in food web structure? *Oecologia*, **74**, 231–235.

Webb, S.D. (1984) On two kinds of rapid faunal turnover, in *Catastrophes and Earth History: the new uniformitarianism*, (eds W.A. Berggren and J.A. Van Couvering), Princeton University Press, Princeton, NJ, pp. 417–436.

Webster, H.O. (1962) Rediscovery of the noisy scrub-bird. *Western Australian Naturalist*, **8**, 57–59.

Werner, E.E. (1974) The fish size, prey size, handling time relation of several sunfishes and some implications. *Journal of the Fisheries Research Board of Canada*, **31**, 1531–1536.

Williams, A.B. (1936) The composition and dynamics of a beech–maple climax community. *Ecological Monographs*, **6**, 318–408.

Wilson, D.S. (1975) The adequacy of body size as a niche difference. *American Naturalist*, **109**, 769–784.

Yodzis, P. (1993) Environment and trophodiversity, in *Species Diversity in Ecological Communities: historical and geographical perspectives*, (eds R. Ricklefs and D. Schluter), University of Chicago Press, Chicago, pp. 26–38.

Zaret, T.M. (1980) *Predation and Freshwater Communities*, Yale University Press, New Haven, CT.

6 Speciation and rarity: separating cause from consequence

Steven L. Chown

6.1 INTRODUCTION

Diversity in any geographical area is a first-order function of immigration, emigration, speciation and extinction (Cracraft, 1986; Archibald, 1993), except at the largest scales where it is an exclusive outcome of differential rates of species production and extinction (Cracraft, 1985). Both ecologists and evolutionary biologists have recognized the importance of these processes, although they have tended to focus their work at either local or more regional scales (Brooks and Wiley, 1988; Ricklefs, 1989). Extinction was of great interest to Darwin, but was of less concern to neoDarwinian theory until recently (Raup, 1994). Almost contemporaneously it was realized that past mass extinctions could have been externally driven (Alvarez *et al.*, 1980), and that humans are precipitating a mass extinction event unparalleled in the earth's history (Ehrlich and Ehrlich, 1981). As a result, both ecologists (see Pimm, 1991) and evolutionary biologists (e.g. Raup, 1992, and almost any recent volume of *Paleobiology*) turned their attention to the processes leading to a decrease in range size and abundance culminating in extinction. Our understanding of the influence of extinction on local, regional and temporal patterns of diversity is consequently progressing at an unprecedented rate.

Despite an almost equally large literature concerning speciation (Otte and Endler, 1989), most of this work has remained concerned with the genetics of speciation (e.g. Coyne and Orr, 1989), modes of speciation, their frequency and likelihood (Mayr, 1963, 1982; Stanley, 1979; Cracraft, 1985; Brooks and McLennan, 1991), and the relationship between species concepts and speciation processes (Vrba, 1985; Chandler and Gromko, 1989). Although these long-standing debates (O'Hara, 1993) have enhanced our understanding of the speciation process, they have also tended

The Biology of Rarity. Edited by William E. Kunin and Kevin J. Gaston.
Published in 1997 by Chapman & Hall, London. ISBN 0 412 63380 9.

to draw attention away from the relationship between species characteristics and speciation. In other words, there is no clear understanding of the relationship between species-level characteristics, the likelihood and mode of speciation, and the consequences thereof for the generation and maintenance of ecological and taxonomic diversity (Kunin and Gaston, 1993). For example:

- Most assemblages are characterized by only a few abundant and often widespread species (Gaston, 1994).
- There is a tendency for the size distribution of species within a lineage to become progressively more right-skewed through time (Stanley, 1973).
- Small species, but not the smallest, predominate in assemblages (Dial and Marzluff, 1988; Blackburn and Gaston, 1994a).
- Most taxa show distributions where usually one taxon is numerically dominant (Dial and Marzluff, 1989; Brooks and McLennan, 1993).

These examples suggest that there may be a relationship between species-level characteristics and speciation rate that deserves further attention. Likewise, it seems likely that speciation will have an effect on species characteristics beyond that of the reduction in abundance and range size that necessarily follows any speciation event (Glazier, 1987).

The aim of this chapter is therefore to provide an exploration of the relationship between speciation and species-level characteristics, particularly abundance, range size, dispersal ability, extent of ecological 'specialization' (Futuyma and Moreno, 1988), and, to a lesser extent, trophic position and body size. More specifically, the subject of investigation will be the notion that the set of rare species could be biased due to differences in speciation rates among taxa with different traits. Although rarity can be defined in many ways (e.g. Rabinowitz, 1981), Gaston's (1994) usage of rarity as the 'state of having a low abundance and/or small range size' will be used. Because much of the information used in this review has its origin in studies not specifically aimed at addressing these questions, the distinction between rare and common species will, however, be less stringent than he proposed.

6.2 SPECIES CONCEPTS AND MODES OF SPECIATION

The wide array of species concepts – e.g. biological species concept (Mayr, 1963), recognition concept of species (Paterson, 1985), evolutionary species concept (Wiley, 1978), phylogenetic species concept (Cracraft, 1986) and cohesion species concept (Templeton, 1989) – and speciation models (Brooks and McLennan, 1991) is liable to bedevil any discussion of the ecology of speciation. There is vigorous debate concerning the validity of these concepts (e.g. Coyne *et al.*, 1988), their applicability to natural systems and the relationship between these concepts and the process of speciation. Many authors suggest that we must know what a species is before we can investigate species formation (Templeton, 1989), and others specifically couple species concepts with modes of speciation (Paterson, 1985). On the other hand, Mishler and Donoghue (1982) argued that species have no special biological reality and that a 'case-by-case' approach be taken to overcome the limitations of various species concepts (see also Sluys, 1991). Furthermore, Chandler and Gromko (1989) suggested that species concepts and speciation processes not only can but should be decoupled in evolutionary studies.

In an incisive analysis of the species problem, O'Hara (1993) argued that all species concepts are dependent on prospective narration, the truth of which is logically indeterminable in the present, and that the species problem is one of historical representation. He suggested that by recognizing the limits of historical representation we will be able to 'get over the species problem'. Although O'Hara's argument is cogent, it seems unlikely to be of great assistance in any analysis of the relationship between species-level characteristics and speciation. In addition, many of the problems associated with the biological, recognition and evolutionary species concepts have been resolved without recourse to either dismissal or pluralism in Templeton's (1989) cohesion concept of species. This concept defines a species as 'the most inclusive group of organisms having the potential for genetic and/or demographic exchangeability' and is applicable to asexual species through to syngameons ('. . . the most inclusive unit of interbreeding in a hybridizing species group').

Brooks and McLennan (1991) provided a thorough review of the classes and modes of speciation, and their terminology will be adhered to throughout. Theoretical studies have recognized at least five modes of speciation. These are:

- allopatric speciation mode I (ASMI, also known as geographical, vicariant or dichopatric speciation);
- allopatric speciation mode II (ASMII, peripheral isolate speciation);
- parapatric speciation;
- alloparapatric speciation;
- sympatric speciation.

There is considerable theoretical and empirical support for ASMI (Cracraft, 1982, 1986; Cracraft and Prum, 1988; Lynch, 1989; M.B. Bush, 1994) and ASMII (Ripley and Beehler, 1990; Enghoff and Baez, 1993; Levin, 1993; Chesser and Zink, 1994; Chown, 1994).

Although both parapatric and sympatric speciation have less empirical support than the allopatric modes, they are nonetheless more common than previously thought. For example, Ripley and Beehler (1990) demonstrated parapatric speciation in Indian birds, and there is a growing body of evidence suggesting that sympatric speciation may be an important process particularly in invertebrates (Rothfels, 1989; Tauber and Tauber, 1989; G.L. Bush, 1994; also Schliewen *et al.*, 1994). Much of the opposition to both parapatric and sympatric speciation was associated with the extremely restrictive nature of earlier theoretical models of these modes. However, in a remarkable review of speciation experiments conducted over the past 40 years, Rice and Hostert (1993) found strong support for the evolution of reproductive isolation via pleiotropy and genetic hitchhiking with or without allopatry. Their analysis of the single variation model of divergence with gene flow showed elegantly that parapatric and sympatric speciation may be common under these conditions, although double variation models remain as restrictive.

Rice and Hostert (1993) concluded that strong divergent selection, sampling drift and sexual/sexually antagonistic selection are the major mechanisms driving speciation, and that geography acts only to restrict gene flow and produce sharp discontinuities in selection. They also showed that allopatry is not a strong pre-requisite for speciation and presented a graphical model demonstrating the relationship between gene flow and 'discontinuous, multifarious, divergent selection', based on speciation via pleiotropy. This model is reproduced in Figure 6.1. When gene flow is large, selection has to be concomitantly strong to promote speciation, but as soon as gene flow is reduced (as in the case of allopatric speciation) weak selection and/or sampling drift is sufficient to ensure divergence. Templeton's (1989) cohesion species concept fits the framework of this model well. For example, at point A in Figure 6.1, speciation can take place despite genetic cohesion. At point B, genetic exchangeability is low but demographic cohesion is high, therefore divergence does not take place. Points C and D represent the more conventional interpretations, i.e. no divergence because of genetic cohesion, and speciation due to selection and an absence of gene flow. Juxtaposing Templeton's (1989) cohesion concept and Rice and Hostert's (1993) pleiotropic speciation model may be of considerable heuristic value when examining the causal relationship between rarity and speciation.

6.3 SPECIES-LEVEL CHARACTERISTICS AND SPECIATION RATE

Before proceeding with an analysis of the relationships between species-level characteristics and speciation rate, it should be noted that the relationships such as those between body size, geographical range size, abundance and dispersal ability *per se* will not be addressed. There is a large literature dealing with these relationships (see Gaston, 1994, for review). In general, there is a positive relationship between range size and abundance, and between range size and dispersal ability, although the relationship between these characteristics and body size is considerably more complex (Blackburn and Gaston, 1994b; Gaston, 1994: 126). The relationships between these characteristics, and the problems associated

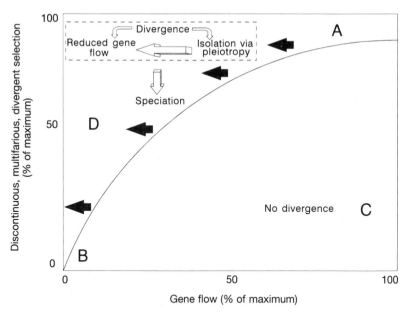

Figure 6.1 Rice and Hostert's (1993) graphical model for speciation via pleiotropy. (Redrawn from *Evolution*, **47**(6).) See text for further explanation.

with interpreting them, seem to hold both for terrestrial and marine taxa (Russell and Lindberg, 1988; Jablonski and Valentine, 1990).

Theoretical studies of the relationship between geographical range size, abundance and speciation have led to two remarkably divergent points of view. Mayr (1963) and Stanley (1979) argued that widespread and abundant taxa, possessing well developed dispersal abilities, should have lower rates of speciation than narrowly distributed taxa with low abundances and poor dispersal ability. Essentially their logic was that gene flow associated with highly vagile taxa is liable to inhibit speciation. On the contrary Vermeij (1987), Snow (1981) and Brooks and McLennan (1993) proposed that taxa with a broad geographical range and a well developed dispersal ability should be highly susceptible to speciation (see also Vrba, 1980, 1987). Vermeij (1987) argued that founder events are favoured by broad geographical range and this line of reasoning was also adopted by Snow (1981). Brooks and McLennan (1993) suggested that an increase in the number of dispersal stages in eucestodes, compared with their sister group, could promote long-term survival and geographical spread of a species, 'increasing the likelihood of vicariant speciation'. Marzluff and Dial (1991) argued that broad geographical range and enhanced colonization ability may actively promote speciation, and that diverse taxa are characterized by short generation times and high mobility which allow rapid colonization, broad geographical ranges and rapid responses to selection. Their ideas were formulated in a graphic model (reproduced in Figure 6.2) and they argued that this model should hold for most taxa. They also suggested that the intrinsic properties of species (i.e. life history traits) directly influence the probability of speciation, and that extrinsic properties are of less importance (*contra* Cracraft, 1985, and many others).

Most evidence with which we can judge these opposing points of view comes from paleoecological and paleobiological studies, many of which have been concerned with the factors influencing speciation and extinction rate (Kitchell, 1985; Allmon, 1994), often as a test of the likelihood of species selection (Stanley, 1979). A number of 'neontological' studies, particularly on plants, are also pertinent (see also Chapter 12). Paleontological studies have demonstrated that in many marine groups those taxa which include species (or sometimes genera) with narrow geographical ranges, low abundances and poor dispersal ability (non-planktonic larvae) have greater speciation (and extinction) rates than taxa where most species are widespread, abundant and have well developed dispersal abilities (planktonic larvae). These groups include bivalves (Jackson, 1974), gastropods (Hansen, 1978; Lieberman *et al.*, 1993), stomatopod crustaceans (Reaka, 1980) and foraminifera (Stanley *et al.*, 1988). Palumbi (1992:

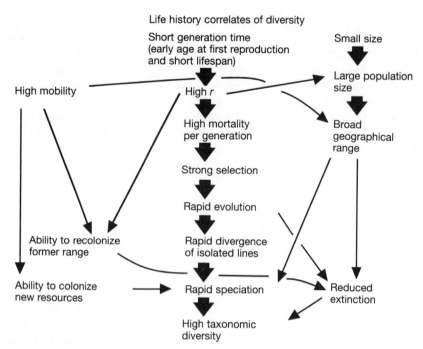

Figure 6.2 Marzluff and Dial's (1991) summary model of the effects of life history and colonization ability on taxonomic diversity. (Redrawn from *Ecology*, **72**(2).)

Table 1) provided evidence from genetic studies in support of these conclusions. Genetic differences among localities tend to be small for species with long planktonic residence times. In their study of the geographical ranges and life-history characteristics of Australian terrestrial plants, Oakwood *et al.* (1993) concluded that lineages with less capacity for long-range dispersal tend to divide up into a large number of species each with a narrow geographical range. Tryon (1970) found that, in ferns, species having more pronounced dispersal abilities are less liable to produce endemics on isolated islands. In the Cape Floristic Region of South Africa, Cowling and Holmes (1992) found that ant-dispersed (rather than vertebrate-dispersed) species were not widespread and possessed a disproportionate number of endemic species. Although there seems to be a large body of evidence supporting the suggestion that small range sizes, low abundance and poor dispersal ability are all associated with an elevated speciation rate, the relationship is not straightforward.

In their studies of Neogene Bivalvia and planktonic foraminifera, Stanley (1986) and Stanley *et al.* (1988) found that the effects of geographical range on speciation are overwhelmed by the effects of population density and are consequently less important in determining speciation rate. Here, small population size and reduced dispersal ability lead to elevated speciation rates. Reduced abundance and patchy and unstable population structure (see also Gaston, 1990) promote speciation.

The picture is further complicated by Vermeij's (1987) finding that widespread marine species tend to be highly susceptible to founder speciation (ASMII) at the edges of their ranges. Rosenzweig (1975) also implied that species with the greatest geographical ranges (perimeters) should shed the most isolates per unit time and thus give rise to most new species. In addition, Palumbi (1992) suggested that in marine species, and terrestrial plants, high dispersal potential does not always lead to high gene flow, and isolation by large distances can result in genetic divergence even in high-dispersal species. Limited evidence for elevated rates of divergence, due to enhanced dispersal abilities, has been found in some angiosperms and ferns (Tiffney, 1984, 1986; Eriksson and Bremer, 1991; Ranker *et al.*, 1994).

Until recently, it was thought that the relationship between vertebrate-assisted dispersal and divergence, in angiosperms at least, is either weak (Eriksson and Bremer, 1991) or absent (Herrera, 1989; Eriksson and Bremer, 1992). A re-analysis of the data for angiosperms by Tiffney and Mazer (1995) revealed that biotic dispersal has enhanced the richness of woody angiosperm taxa possessing this trait, mostly via a reduction in extinction probability. In contrast, elevated rates of diversification in herbaceous angiosperms are associated with abiotic dispersal (lower dispersal ability). In the latter case, however, elevated rates of diversification are also thought to be a consequence of narrower ecological tolerances, or specialization.

Although specialization cannot be broadly defined (Vrba, 1987; Futuyma and Moreno, 1988), there is considerable evidence that special- ization along one or more niche axes is responsible for elevated speciation rates. Stanley (1986) concluded that, in Neogene bivalves, increasing behavioural complexity and stenotopy (restricted habitat choice) tend to enhance speciation rates. Baumiller (1993) came to a similar conclusion with regard to Paleozoic Crinoidea. Vrba (1980, 1987) demonstrated that in African mammals (mostly Artiodactyla) clades of specialist feeders have higher speciation rates than those with less specialized feeding habits. In 11 of 13 insect sister-group comparisons, Mitter *et al.* (1988) showed that the phytophagous lineage was considerably more diverse than the non- phytophagous one. Farrell *et al.* (1991) demonstrated that in 13 of 16 plant sister-group comparisons, the lineage bearing secretory canals is more diverse than its presumed sister group. They argued that the specialization (escape from herbivory) promoted diversification.

On the contrary, Wiegmann *et al.* (1993) found that, in insects, specialization as carnivorous parasites does not enhance diversification but may, in fact, retard it. They argued that specialization *per se* may not be responsible for elevated speciation rates, and that resource quality and availability may have determined the scope for elevated speciation rates. Vrba (1987) also failed to demonstrate enhanced diversification in a higher trophic level, specialist consumer (the myrmecophagous mammalian genus *Orycteropus*). She argued that generalists, as well as clades whose narrow resources tend to persist even as their larger environments change (i.e. persistent-patch specialists), should show low speciation and extinc- tion rates. It seems likely that many secondary consumers, and taxa in higher trophic levels, such as parasitic insects, will reflect this pattern. However, due to the preponderance of primary producers and consumers in all ecosystems, the overriding pattern is likely to be one where elevated diversification rates are associated with increasing specialization.

The relationship between body size, specialization and speciation rate is considerably more complex. Lindström *et al.* (1994) and Wasserman and Mitter (1978) showed that, in Holarctic moths, polyphagous species tend to have larger body sizes than oligophages or monophages. In their investigation of body size and species composition of North American land mammals, Brown and Nicoletto (1991) argued that larger mammal species tend to have greater vagility and broader environmental tolerances, which would result in lower speciation rates in these compared with smaller species. Brown *et al.* (1993) argued that there is an optimal size for mammals (and other taxa). In a particular habitat the first 'invading' species should assume the optimal size, with a spread of sizes developing around this optimum (Blackburn and Gaston, 1994b). This micro- evolutionary process (selection) could lead, in conjunction with elevated rates of cladogenesis in smaller-sized taxa, to the right-skewed distribution of body sizes in many taxa (Maurer *et al.*, 1992). Thus there appear to be some grounds for arguing that, within specific higher taxa, elevated speciation rates are associated with relatively small body sizes. However, this relationship is unlikely to be straightforward. For example, Blackburn and Gaston (1996) have argued that there may not be an optimal body size for higher taxa. In addition, abundance is thought to decrease and geographical range size increase with increasing body size in many taxa (the debate is complex and ongoing: for a review see e.g. Currie and Fritz, 1993; Gaston, 1994), while speciation rate decreases with increasing geographical range size and abundance (see above). It is not clear how these factors interact to either elevate or depress speciation rate, although a number of recent paleontological studies provide some insight regarding these issues.

In his discussion of the factors that accelerate both speciation and extinction, Stanley (1986) suggested that hostile environmental conditions on a broad geographical scale markedly influence both processes, and labelled this the 'fission effect'. Stanley outlined a graphical model of this effect for burrowing bivalves (Figure 6.3), which shows that at very small population sizes extinction rate is high and speciation rate low. As population size increases, there is a curvilinear decline in extinction rate and an almost exponential increase in speciation rate. However, the speciation rate increases only up to a point and declines to a considerably lower level thereafter. If it is presumed that the same relationship holds for both geographical range and dispersal ability, which seems likely (see above), the abscissa should be accordingly relabelled. In addition, if the model is viewed from a single species perspective, a more appropriate

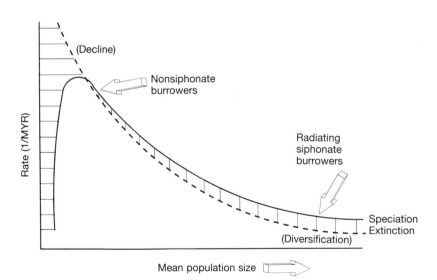

Figure 6.3 Stanley's (1986) model of the 'fission effect' for Neogene bivalves. (Redrawn from *Paleobiology*, **12**(1).)

label for the ordinate is probability. If the axes are modified accordingly and considered in light of Rice and Hostert's (1993) model of speciation and Templeton's (1989) cohesion species concept, a graphical model exploring causal relationships between range size, abundance, dispersal and rate of speciation (and extinction) emerges. This modified model is shown in Figure 6.4.

Here, geographical range size, abundance and dispersal ability are presumed to be positively interrelated. However, this is not a prerequisite, nor does independence of these characteristics weaken the value of the model. The area marked A represents narrow geographical range and low abundance. Species in this area are fated with extinction for various reasons (Pimm, 1991; Lande, 1993). At this stage, high dispersal rates would simply represent emigration, adding to the chances of extinction. Therefore, the 'extinction curve' may not be entirely apposite with regard to this characteristic, except from a 'rescue effect' (Pimm, 1991) point of view. Likewise, genetic and demographic cohesion in such a small population is liable either to be so high as to eliminate the chances of speciation, or non-existent, thus interfering with reproduction (in sexually reproducing species). As abundance, range size and dispersal ability increase, extinction rate declines due to the rescue effect from other populations (Pimm, 1991) and the distribution of populations over a wider geographical range, which facilitates escape from inimical environmental conditions (Kitchell, 1985). At the same time, the distribution of populations over a wider geographical area increases the likelihood of discontinuous, multifarious, divergent selection at a rate faster than that

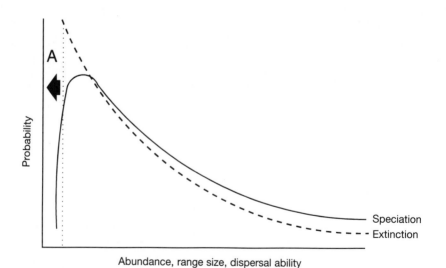

Figure 6.4 Modification of Stanley's (1986) 'fission effect' model to a single species 'rarity effect' model. (See text for explanation.)

which can be overcome by gene-flow via dispersal (Figure 6.1), resulting in an increased probability of speciation (see also Chapter 12). Factors such as homing ability would act to increase this differential rate of selection intensity and gene flow. As dispersal ability continues to rise and populations become more panmictic, the probability of speciation declines rapidly. However, speciation probability remains higher than that of extinction due to the possibility of vicariance, resulting from extrinsic factors and/or the formation of peripheral isolates.

The effects of body size on range size and dispersal ability appear to be quite straightforward in this model, but the effects of the nonlinear relationship between body size, abundance and speciation probability are less clear and remain to be explored. Nonetheless, it seems likely that the model will hold for allopatric, parapatric and sympatric modes of speciation. In the latter case, geographical range is of less concern than factors such as mating sites, host specialization, etc. (but see Chapter 12). In addition, the interrelationship described here between speciation rate and body size, resource acquisition and conversion, abundance and range size may provide an evolutionary ecological explanation for high tropical diversity (Stevens, 1989) and the tropical origin of many taxa (Jablonski, 1993), which takes account of both intrinsic and extrinsic factors.

With regard to rarity it should be clear that rare species, but not the rarest species, have a higher speciation probability than more common species, and that these species are liable to be small and specialized along one or more niche axes relative to more common species. However, this pattern is unlikely to be consistent between trophic levels, due to an increase in the number of persistent-patch specialists at higher trophic levels. In addition, it should be noted that those factors which promote speciation are precisely those which are responsible for an enhanced probability of extinction in many groups (Kitchell, 1985; Masters and Rayner, 1993; Vrba, 1993). This has considerable implications with regard to rarity; high turnover taxa may necessarily be rare (section 6.5).

6.4 RARITY AS A CONSEQUENCE OF SPECIATION

A reduction in range size and abundance is a consequence of any speciation event (Glazier, 1987). Stanley (1986) pointed out that nothing should prevent the dispersal of new species, but that they will normally also be characterized by low densities. This seems to be particularly so if range size and abundance are heritable species-level characteristics (Jablonski *et al.*, 1985; Jablonski, 1987). Not all taxa can be subject to this process, hence the fact that common and widespread taxa are found (see also Stanley, 1973). Nonetheless, through evolutionary time, clades possibly do fall prey to a process which leads to ongoing stenotopy, behavioural complexity, specialization, rarity and eventual extinction during times of environmental change (e.g. Stanley, 1979; Lieberman *et al.*, 1993). Such canalization (Brooks and Wiley, 1988) could then only be avoided via heterochrony (changes in timing of development) and a return to a more generalist groundplan (Holm, 1985; Futuyma, 1986). However, there is some evidence, particularly for ASMII, that speciation can involve an expansion and subsequent contraction of range size and abundance. This point was mooted by Lawton (1993) in his discussion of range, population abundance and conservation and has, in fact, been the subject of considerable research.

In an analysis of the evolution of Melanesian ant faunas, Wilson (1959, 1961) suggested that insular species move through a series of stages from newly arrived colonists to highly differentiated endemic species that eventually become extinct. This taxon cycle has been documented for a variety of other island taxa (Greenslade, 1968; Ricklefs and Cox, 1978; Rummel and Roughgarden, 1985) and continental taxa (Glazier, 1980; Erwin, 1985). In Stage I of the cycle, a species colonizes an archipelago and expands its range, usually occupying most or all of the islands. In other words, the species moves from having a low abundance and narrow range to being abundant and widespread. In Stage II, populations on the islands have differentiated sufficiently to be recognized as endemic subspecies, and the species may have gone extinct on a few islands. In Stage III, most populations are sufficiently different to be recognized as endemic species and many populations are extinct. Finally, in Stage IV, a single population remains as an endemic species on a single island. In other words range size and abundance are considerably reduced.

As species move through this cycle, or pulse, their life history characteristics evolve from those of generalized species inhabiting disturbed habitats (Wilson, 1961; Glazier, 1980, 1987) to those of more specialized species inhabiting more stable habitats. Glazier (1987) suggested that, since small populations with narrow geographical range sizes are more susceptible to extinction, any ecological characteristic that confers enhanced survivorship on such isolates is expected to be preferentially represented. The use of resources and habitats that confer stability are undoubtedly such characteristics. This may explain the passage of isolates through the cycle. Alternative explanations invoke competition between invaders and residents (Ricklefs and Cox, 1978; Rummel and Rough-garden, 1985) and the differential effects of climate change on habitat generalists and specialists (Pregill and Olson, 1981; see also Chown, 1994).

Although the likelihood of taxon cycles has been questioned (*Anolis* lizards: Losos, 1992; carabid beetles: Liebherr and Hajek, 1990), taxon cycles remain theoretically possible on all but isolated islands (Taper and Case, 1992), and there is persistent empirical evidence that they do occur on both continents and islands (Roughgarden and Pacala, 1989; Liebherr and Hajek, 1990; Roughgarden, 1992). This change in abundance and range size for single species moving through the 'cycle' or pulse eventually leads to a reduction in both range size and abundance. Glazier (1980) elegantly demonstrated this for continental *Peromyscus* species, providing further evidence for the tendency of common, widespread species to give rise to species with range sizes and abundances smaller than those expected solely from lineage splitting. Taxon cycles therefore present an excellent opportunity for studying the consequences of speciation on range size and abundance and should be further explored using more modern techniques (Lawton, 1993).

6.5 CONCLUSIONS

In the Introduction it was suggested that some causal relationship between species-level characteristics and speciation may be responsible for the non-random patterns that have been documented in ecological and taxonomic assemblages. It now seems clear that these patterns are partially a result of differential speciation rates associated with commonness and rarity, although this relationship is not straightforward, as the 'rarity effects' model demonstrates. Rarity may also be a regular outcome of speciation. One of the consequences of this two-way interaction between rarity and speciation is that taxa may end up in a 'trap', ultimately leading to an evolutionary dead-end and extinction. Nonetheless, the way in which the relationship between speciation and rarity varies between taxa of different body sizes, and from different habitats and trophic levels, is far from clear. Although paleoecology has produced a flood of information on and insight into these relationships (Kitchell, 1985), our understanding of the interaction between rarity and speciation depends not only on the continuation of such work but also on the expansion of evolutionary ecological approaches to the problem. A combination of phylogenetic analysis and community and population ecology is liable to be most useful in this regard (Brooks *et al.*, 1992; Archibald, 1993). Phylogenetic studies are essential for determining sister taxa, which are by definition of the same age. Subsequent studies of the life history characteristics of member species within these sister taxa, and their relative abundance and distribution, with regard both to the ecological and taxonomic assemblages they are part of, may provide considerable insight into the relationship between speciation and rarity.

Perhaps one of the most interesting and concerning conclusions of this study is that the characteristics associated with elevated speciation rates are also those associated with increased extinction rates. This may be a result of reasonably straightforward relationships between species characteristics, and extinction and speciation probabilities, but it may also be a consequence of the fact that self-organization, in conjunction with selection, may link speciation, extinction and ecological web structure (Kauffman, 1993: 277). Whatever the reasons for this relationship, it is of considerable concern given the rapid fragmentation and destruction of habitat on a global scale (Soulé, 1991). If relatively rare species are those that are most likely to suffer extinction (Lawton, 1993; Gaston, 1994; and above) but are also those that are most likely to give rise to new species, the capacity for recovery following this human-induced mass extinction is liable to be small (Stanley, 1990). In addition, if Kauffman's (1993) theoretical analysis has any support, we should expect that, as habitat destruction and fragmentation force species into region A of the rarity effects model (Figure 6.4), the likelihood of a rapid 'melting' of the ecological web and global ecosystem instability will increase dramatically.

ACKNOWLEDGEMENTS

M.E. Pienaar assisted with the compilation of literature. T.J. Robinson and M.A. McGeoch (University of Pretoria) reviewed the manuscript. T.M. Blackburn and K.J. Gaston (Imperial College) are thanked for providing an unpublished manuscript. This work was supported by the University of Pretoria and the Foundation for Research Development. The referees and W.E. Kunin and K.J. Gaston are thanked for their constructive criticism.

REFERENCES

Allmon, W.D. (1994) Taxic evolutionary paleoecology and the ecological context of macro-evolutionary change. *Evolutionary Ecology*, **8**, 95–112.

Alvarez, L.W., Alvarez, W., Asaro, F. and Michel, A.V. (1980) Extraterrestrial causes for the Cretaceous–Tertiary extinction. *Science*, **208**, 1095–1108.

Archibald, J.D. (1993) The importance of phylogenetic analysis for the assessment of species turnover: a case history of Paleocene mammals in North America. *Paleobiology*, **19**, 1–27.

Baumiller, T.K. (1993) Survivorship analysis of Paleozoic Crinoidea: effect of filter morphology on evolutionary rates. *Paleobiology*, **19**, 304–321.

Blackburn, T.M. and Gaston, K.J. (1994a) The distribution of body sizes of the world's bird species. *Oikos*, **70**, 127–130.

Blackburn, T.M. and Gaston, K.J. (1994b) Animal body size distributions: patterns, mechanisms and implications. *Trends in Ecology and Evolution*, **9**, 471–474.

Blackburn, T.M. and Gaston, K.J. (1996) On being the right size: different definitions of 'right'. *Oikos*, **75**, 551–557.

Brooks, D.R. and McLennan, D.A. (1991) *Phylogeny, Ecology, and Behaviour. A Research Programme in Comparative Biology*, The University of Chicago Press, Chicago.

Brooks, D.R. and McLennan, D.A. (1993) Comparative study of adaptive radiations with an example using parasitic flatworms (Platyhelminthes: Cercomeria). *American Naturalist*, **142**, 755–778.

Brooks, D.R. and Wiley, E.O. (1988) *Evolution as Entropy. Toward a Unified Theory of Biology*, 2nd edn, The University of Chicago Press, Chicago.

Brooks, D.R., Mayden, R.L. and McLennan, D.A. (1992) Phylogeny and biodiversity: Conserving our evolutionary legacy. *Trends in Ecology and Evolution*, **7**, 55–59.

Brown, J.H. and Nicoletto, P.F. (1991) Spatial scaling of species composition: body masses of North American land mammals. *American Naturalist*, **138**, 1478–1512.

Brown, J.H., Marquet, P.A. and Taper, M.L. (1993) Evolution of body size: consequences of an energetic definition of fitness. *American Naturalist*, **142**, 573–584.

Bush, G.L. (1994) Sympatric speciation in animals: new wine in old bottles. *Trends in Ecology and Evolution*, **9**, 285–288.

Bush, M.B. (1994) Amazonian speciation: a necessarily complex model. *Journal of Biogeography*, **21**, 5–17.

Chandler, C.R. and Gromko, M.H. (1989) On the relationship between species concepts and speciation processes. *Systematic Zoology*, **38**, 116–125.

Chesser, R.T. and Zink, R.M. (1994) Modes of speciation in birds: A test of Lynch's method. *Evolution*, **48**, 490–497.

Chown, S.L. (1994) Historical ecology of subantarctic weevils: patterns and processes on isolated islands. *Journal of Natural History*, **28**, 411–433.

Cowling, R.M. and Holmes, P.M. (1992) Endemism and speciation in a lowland flora from the Cape Floristic Region. *Biological Journal of the Linnean Society*, **47**, 367–383.

Coyne, J.A. and Orr, H.A. (1989) Patterns of speciation in *Drosophila*. *Evolution*, **43**, 362–381.

Coyne, J.A., Orr, H.A. and Futuyma, D.J. (1988) Do we need a new species concept? *Systematic Zoology*, **37**, 190–200.

Cracraft, J. (1982) Geographic differentiation, cladistics, and vicariance biogeography: reconstructing the tempo and mode of evolution. *American Zoologist*, **22**, 411–424.

Cracraft, J. (1985) Biological diversification and its causes. *Annals of the Missouri Botanical Garden*, **72**, 794–822.

Cracraft, J. (1986) Origin and evolution of continental biotas: speciation and historical congruence within the Australian avifauna. *Evolution*, **40**, 977–996.

Cracraft, J. and Prum, R.O. (1988) Patterns and processes of diversification: speciation and historical congruence in some neotropical birds. *Evolution*, **42**, 603–620.

Currie, D.J. and Fritz, J.T. (1993) Global patterns of animal abundance and species energy use. *Oikos*, **67**, 56–68.

Dial, K.P. and Marzluff, J.M. (1988) Are the smallest organisms the most diverse? *Ecology*, **69**, 1620–1624.

Dial, K.P. and Marzluff, J.M. (1989) Nonrandom diversification within taxonomic assemblages. *Systematic Zoology*, **38**, 26–37.

Ehrlich, P.R. and Ehrlich, A.H. (1981). *Extinction. The Causes and Consequences of the Disappearance of Species*, Random House, New York.

Enghoff, H. and Baez, M. (1993) Evolution of distribution and habitat patterns in endemic millipedes of the genus *Dolichoiulus* (Diplopoda: Julidae) on the Canary Islands, with notes on distribution patterns of other Canarian species swarms. *Biological Journal of the Linnean Society*, **49**, 277–301.

Eriksson, O. and Bremer, B. (1991) Fruit characteristics, life forms, and species richness in the plant family Rubiaceae. *American Naturalist*, **138**, 751–761.

Eriksson, O. and Bremer, B. (1992) Pollination systems, dispersal modes, life forms, and diversification rates in angiosperm families. *Evolution*, **46**, 258–266.

Erwin, T.L. (1985) The taxon pulse: a general pattern of lineage radiation and extinction among carabid beetles, in *Taxonomy, Phylogeny and Zoogeography of Beetles and Ants* (ed. G.E. Ball), Dr W. Junk, Dordrecht, pp. 437–472.

Farrell, B.D., Dussourd, D.E. and Mitter, C. (1991) Escalation of plant defense: do latex and resin canals spur plant diversification? *American Naturalist*, **138**, 881–900.

Futuyma, D.J. (1986) *Evolutionary Biology*, 2nd edn, Sinauer Associates, Sunderland.

Futuyma, D.J. and Moreno, G. (1988) The evolution of ecological specialization. *Annual Review of Ecology and Systematics*, **19**, 207–233.

Gaston, K.J. (1990) Patterns in the geographical ranges of species. *Biological Reviews*, **65**, 105–129.

Gaston, K.J. (1994) *Rarity*, Chapman & Hall, London.

Glazier, D.S. (1980) Ecological shifts and the evolution of geographically restricted species of North American *Peromyscus* (mice). *Journal of Biogeography*, **7**, 63–83.

Glazier, D.S. (1987) Toward a predictive theory of speciation: the ecology of isolate selection. *Journal of Theoretical Biology*, **126**, 323–333.

Greenslade, P.J.M. (1968) The distribution of some insects of the Solomon Islands. *Proceedings of the Linnean Society of London*, **179**, 189–196.

Hansen, T.A. (1978) Larval dispersal and species longevity in lower tertiary gastropods. *Science*, **199**, 885–887.

Herrera, C.M. (1989) Seed dispersal by animals: a role in angiosperm diversification? *American Naturalist*, **133**, 309–322.

Holm, E. (1985) The evolution of generalist and specialist taxa, in *Species and Speciation* (ed E.S. Vrba), Transvaal Museum Monograph No. 4, Pretoria, pp. 87–93.

Jablonski, D. (1987) Heritability at the species level: analysis of geographic ranges of Cretaceous molluscs. *Science*, **238**, 360–363.

Jablonski, D. (1993) The tropics as a source of evolutionary novelty through geological time. *Nature*, **364**, 142–144.

Jablonski, D. and Valentine, J.W. (1990) From regional to total geographic ranges: testing the relationship in Recent bivalves. *Paleobiology*, **16**, 126–142.

Jablonski, D., Flessa, K.W. and Valentine, J.W. (1985) Biogeography and paleobiology. *Paleobiology*, **11**, 75–90.

Jackson, J.B.C. (1974) Biogeographic consequences of eurytopy and stenotopy among marine bivalves and their evolutionary significance. *American Naturalist*, **108**, 541–560.

Kauffman, S.A. (1993) *The Origins of Order. Self-Organization and Selection in Evolution*, Oxford University Press, Oxford.

Kitchell, J.A. (1985) Evolutionary paleoecology: recent contributions to evolutionary theory. *Paleobiology*, **11**, 91–104.

Kunin, W.E. and Gaston, K.J. (1993) The biology of rarity: patterns, causes and consequences. *Trends in Ecology and Evolution*, **8**, 208–301.

Lande, R. (1993) Risks of population extinction from demographic and environmental stochasticity and random catastrophes. *American Naturalist*, **142**, 911–927.

Lawton, J. (1993) Range, population abundance and conservation. *Trends in Ecology and Evolution*, **8**, 409–413.

Levin, D.A. (1993) Local speciation in plants: the rule not the exception. *Systematic Botany*, **18**, 197–208.

Lieberman, B.S., Warren, D.A. and Eldredge, N. (1993) Levels of selection and macro-evolutionary patterns in the turritellid gastropods. *Paleobiology*, **19**, 205–215.

Liebherr, J.K. and Hajek, A.E. (1990) A cladistic test of the taxon cycle and taxon pulse hypotheses. *Cladistics*, **6**, 39–59.

Lindström, J., Kaila, L. and Niemelä, P. (1994) Polyphagy and adult body size in geometrid moths. *Oecologia*, **98**, 130–132.

Losos, J.B. (1992) A critical comparison of the taxon-cycle and character-displacement models for size evolution of *Anolis* Lizards in the Lesser Antilles. *Copeia*, **1992**, 279–288.

Lynch, J.D. (1989) The gauge of speciation: on the frequencies of modes of speciation, in *Speciation and its Consequences*, (eds D. Otte and J.A. Endler), Sinauer Associates, Sunderland, pp. 527–553.

Marzluff, J.M. and Dial, K.P. (1991) Life history correlates of taxonomic diversity. *Ecology*, **72**, 428–439.

Masters, J.C. and Rayner, R.J. (1993) Competition and macroevolution: the ghost of competition yet to come? *Biological Journal of the Linnean Society*, **49**, 87–98.

Maurer, B.A., Brown, J.H. and Rusler, R.D. (1992) The micro and macro in body size evolution. *Evolution*, **46**, 939–953.

Mayr, E. (1963) *Animal Species and Evolution*, Harvard University Press, Cambridge, MA.

Mayr, E. (1982) Speciation and macroevolution. *Evolution*, **36**, 1119–1132.

Mishler, B.D. and Donoghue, M.J. (1982) Species concepts: a case for pluralism. *Systematic Zoology*, **31**, 491–503.

Mitter, C., Farrell, B. and Wiegmann, B. (1988) The phylogenetic study of adaptive zones: has phytophagy promoted insect diversification? *American Naturalist*, **132**, 107–128.

Oakwood, M., Jurado, E., Leishman, M. and Westoby, M. (1993) Geographic ranges of plant species in relation to dispersal morphology, growth form and diaspore weight. *Journal of Biogeography*, **20**, 563–572.

O'Hara, R.J. (1993) Systematic generalization, historical fate, and the species problem. *Systematic Biology*, **42**, 231–246.

Otte, D. and Endler, J.A. (eds) (1989) *Speciation and its Consequences*, Sinauer Associates, Sunderland.

Palumbi, S.R. (1992) Marine speciation on a small planet. *Trends in Ecology and Evolution*, **7**, 114–118.

Paterson, H.E.H. (1985) The recognition concept of species, in *Species and Speciation* (ed. E.S. Vrba), Transvaal Museum Monograph No. 4, Pretoria, pp. 21–29.

Pimm, S.L. (1991) *The Balance of Nature? Ecological Issues in the Conservation of Species and Communities*, The University of Chicago Press, Chicago.

Pregill, G.K. and Olson, S.L. (1981) Zoogeography of West Indian vertebrates in relation to Pleistocene climatic cycles. *Annual Review of Ecology and Systematics*, **12**, 75–98.

Rabinowitz, D. (1981) Seven forms of rarity, in *The Biological Aspects of Rare*

Plant Conservation (ed H. Synge), John Wiley and Sons, Chichester, pp.189–217.

Ranker, T.A., Floyd, S.K., Windham, M.D. and Trapp, P.G. (1994) Historical biogeography of *Asplenium adiantum-nigrum* (Aspleniaceae) in North America and implications for speciation theory in homosporous pteridophytes. *American Journal of Botany*, **81**, 776–781.

Raup, D.M. (1992) Large-body impact and extinction in the Phanerozoic. *Paleobiology*, **18**, 80–88.

Raup, D.M. (1994) The role of extinction in evolution. *Proceedings of the National Academy of Science of the USA*, **91**, 6758–6763.

Reaka, M.L. (1980) Geographic range, life history patterns, and body size in a guild of coral-dwelling mantis shrimps. *Evolution*, **34**, 1019–1030.

Rice, W.R. and Hostert, E.E. (1993) Laboratory experiments on speciation: what have we learned in 40 years? *Evolution*, **47**, 1637–1653.

Ricklefs, R.E. (1989) Speciation and diversity: The integration of local and regional processes, in *Speciation and its Consequences*, (eds D. Otte and J.A. Endler), Sinauer Associates, Sunderland, pp. 599–622.

Ricklefs, R.E. and Cox, G.W. (1978) Stage of taxon cycle, habitat distribution and population density in the avifauna of the West Indies. *American Naturalist*, **122**, 875–895.

Ripley, S.D. and Beehler, B.M. (1990) Patterns of speciation in Indian birds. *Journal of Biogeography*, **17**, 639–648.

Rosenzweig, M.L. (1975) On continental steady states of species diversity, in *Ecology and Evolution of Communities* (eds M.L. Cody and J.M. Diamond), Harvard University Press, Cambridge MA, pp. 121–140.

Rothfels, K. (1989) Speciation in black flies. *Genome*, **32**, 500–509.

Roughgarden, J. (1992) Comments on the paper by Losos: character displacement versus taxon loop. *Copeia*, **1992**, 288–295.

Roughgarden, J. and Pacala, S. (1989) Taxon cycle among *Anolis* lizard populations: review of evidence, in *Speciation and its Consequences*, (eds D. Otte and J.A. Endler), Sinauer Associates, Sunderland, pp. 403–432.

Rummel, J.D. and Roughgarden, J. (1985) A theory of faunal buildup for competition communities. *Evolution*, **39**, 1009–1033.

Russell, M.P. and Lindberg, D.R. (1988) Real and random patterns associated with molluscan spatial and temporal distribution. *Paleobiology*, **14**, 322–330.

Schliewen, U.K., Tautz, D. and Paabo, S. (1994) Sympatric speciation suggested by monophyly of crater lake cichlids. *Nature*, **368**, 629–632.

Sluys, R. (1991) Species concepts, process analysis, and the hierarchy of nature. *Experientia*, **47**, 1162–1170.

Snow, D.W. (1981) Tropical frugivorous birds and their food plants: a world survey. *Biotropica*, **13**, 1–14.

Soulé, M.E. (1991) Conservation: Tactics for a constant crisis. *Science,* **253**, 744–750.

Stanley, S.M. (1973) An explanation for Cope's rule. *Evolution*, **27**, 1–26.

Stanley, S.M. (1979) *Macroevolution. Pattern and Process*, W.H. Freeman, San Francisco.

Stanley, S.M. (1986) Population size, extinction, and speciation: the fission effect in Neogene Bivalvia. *Paleobiology*, **12**, 89–110.

Stanley, S.M. (1990) Delayed recovery and the spacing of major extinctions. *Paleobiology*, **16**, 401–414.

Stanley, S.M., Wetmore, K.L. and Kennett, J.P. (1988) Macroevolutionary differences between the two major clades of Neogene planktonic foraminifera. *Paleobiology*, **14**, 235–249.

Stevens, G.C. (1989) The latitudinal gradient in geographical range: how so many species co-exist in the tropics. *American Naturalist*, **133**, 240–256.

Taper, M.L. and Case, T.J. (1992) Models of character displacement and the theoretical robustness of taxon cycles. *Evolution*, **46**, 317–333.

Tauber, C.A. and Tauber, M.J. (1989) Sympatric speciation in insects: perception and perspective, in *Speciation and its Consequences*, (eds D. Otte and J.A. Endler), Sinauer Associates, Sunderland, pp. 307–344.

Templeton, A.R. (1989) The meaning of species and speciation: a genetic perspective, in *Speciation and its Consequences*, (eds D. Otte and J.A. Endler), Sinauer Associates, Sunderland, pp. 3–27.

Tiffney, B.H. (1984) Seed size, dispersal syndromes, and the rise of the angiosperms: evidence and hypothesis. *Annals of the Missouri Botanical Garden*, **71**, 551–576.

Tiffney, B.H. (1986) Fruit and seed dispersal and the evolution of the Hamamelidae. *Annals of the Missouri Botanical Garden*, **73**, 394–416.

Tiffney, B.H. and Mazer, S.J. (1995) Angiosperm growth habit, dispersal and diversification reconsidered. *Evolutionary Ecology*, **9**, 93–117.

Tryon, R. (1970) Development and evolution of fern floras of oceanic islands. *Biotropica*, **2**, 76–84.

Vermeij, G.J. (1987) The dispersal barrier in the tropical Pacific: implications for molluscan speciation and extinction. *Evolution*, **41**, 1046–1058.

Vrba, E.S. (1980) Evolution, species and fossils: How does life evolve? *South African Journal of Science*, **76**, 61–84.

Vrba, E.S. (1985) Environment and evolution: alternative causes of the temporal distribution of evolutionary events. *South African Journal of Science*, **81**, 229–236.

Vrba, E.S. (1987) Ecology in relation to speciation rates: some case histories of Miocene – Recent mammal clades. *Evolutionary Ecology*, **1**, 283–300.

Vrba, E.S. (1993) Turnover-pulses, the red queen, and related topics. *American Journal of Science*, **293**, 418–452.

Wasserman, S.S. and Mitter, C. (1978) The relationship of body size to breadth of diet in some Lepidoptera. *Ecological Entomology*, **3**, 155–160.

Wiegmann, B.M., Mitter, C. and Farrell, B. (1993) Diversification of carnivorous parasitic insects: extraordinary radiation or specialized dead end? *American Naturalist*, **142**, 737–754.

Wiley, E.O. (1978) The evolutionary species concept revisited. *Systematic Zoology*, **27**, 17–26.

Wilson, E.O. (1959) Adaptive shift and dispersal in a tropical ant fauna. *Evolution*, **13**, 122–144.

Wilson, E.O. (1961) The nature of the taxon cycle in the Melanesian ant fauna. *American Naturalist*, **95**, 169–193.

7 How do rare species avoid extinction? A paleontological view

Michael L. McKinney

7.1 INTRODUCTION

One of the ways that rare species may qualitatively differ from common species is through extinction (exit) biases. The set of rare species may be biased by the selective elimination of species that cannot persist at low abundances (Kunin and Gaston, 1993). This chapter reviews the large paleontological literature on extinction for evidence that some rare species are more prone to extinction than others and for evidence about which traits promote such extinction-proneness. It will show that there is fossil evidence about both aspects. This evidence indicates that the set of rare species (however defined) is indeed biased in favour of those species with traits that promote species longevity by being resistant and/or resilient to disturbances. Traits that will be seen as promoting species longevity, and could operate in rare species, include widespread geographical range (for species that are locally sparse), wide niche breadth, morphological and behavioural simplicity, detritivory (and a suite of other traits in marine organisms) and small body size. The evidence for these traits is, as yet, only suggestive because of sampling problems in the fossil record, especially with rare species. But growing paleontological interest in extinction selectivity, especially at finer taxonomic scales, will provide ways to refine the evidence discussed here.

7.2 RARE SPECIES IN THE FOSSIL RECORD

The fossil record is the only place to study most species and most extinctions. Perhaps 5–50 billion species have existed on earth (Raup, 1991). Given, at most, an estimated 50 million species alive today (May, 1988), then well over 99% of earth's species have been extinct since long

The Biology of Rarity. Edited by William E. Kunin and Kevin J. Gaston.
Published in 1997 by Chapman & Hall, London. ISBN 0 412 63380 9.

before humans evolved. In addition, the fossil record is the only place to examine 'natural' extinction dynamics, without complications from human disturbances. It is also the only way of directly examining dynamics over long time periods, at scales ranging from hundreds to millions of years.

A main disadvantage of fossil data is the well known incompleteness of the record. Estimates of the percentage of species leaving a fossil record range from fewer than one to a few per cent (Raup, 1994; Sepkoski, 1994). Furthermore, the incompleteness is highly non-random, with some groups having a much better record than others. The resulting distortion is both environmental and taxonomic, at many spatial scales. Marine taxa thus have a much better record than most of the biosphere but, within the category of all marine taxa, some taxa have much better records than others. There are many reasons why some groups have better fossil records than others (Allison and Briggs, 1991). Environmental biases arise because some habitats, such as marine sediment, are much better environments for deposition and preservation than others. Taxonomic biases arise because some taxa have more durable hard parts, certain life habits, abundant populations or other biological traits that increase the probability of fossilization of many individuals.

Of special relevance to this book are the various biases that affect fossil abundance. Species with greater abundance, however measured, are more likely to be preserved and discovered as fossils. Thus, on average, a species with low local abundance and/or restricted geographical range, is less likely to be preserved or discovered as a fossil than those with the opposite traits. Preservational biases often convert an abundant species into a rare fossil, but the reverse, to convert a rare species into an abundant fossil, is less likely (McKinney and Allmon, 1995).

As most species are not abundant (Gaston, 1994), this bias has profound implications for previous work on the fossil record which is thus largely based on the dynamics of abundant species. Raup (1988), for example, notes that his fossil extinction rate estimates probably underestimate true extinction rates by a factor of 10 because they omit endemic species. Similarly, the apparent 'stasis' of paleocommunities for millions of years in the fossil record may simply occur because the many rare species with high turnover in those communities leave little or no fossil record (McKinney *et al.*, in press).

Sepkoski (1994) reviews efforts to quantify this fossil bias toward abundant species. Figure 7.1, based on his estimates, shows that only a tiny fraction of species present in the fossil record has been sampled. With random sampling, 90% of sampled individuals come from just 0.04% of the total fossil species. This 0.04% represents the most abundant 0.5% of the sampled species. Even when sampling is biased towards rare species, there is still an enormous omission of most rare species (Figure 7.1). These estimates are almost astonishing when one considers that they include only the proportion of actually fossilized species that have been sampled. Given

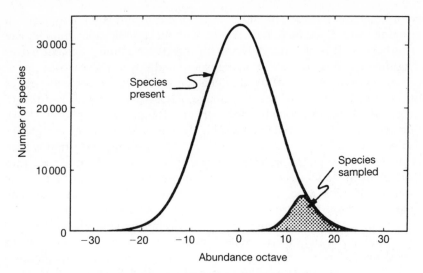

Figure 7.1 According to models of random sampling (Sepkoski, 1994), 40 226 fossil species described from the Cenozoic Era mainly represent the most abundant species in the total fossil pool of 604 650 Cenozoic species that were preserved. Abundance 'octaves' are calculated on the empirical evidence that abundance in both fossil and living species is distributed in log-normal fashion (Sepkoski, 1994). This is also predicted by theory, which indicates that log-normal distributions are common in heterogenous assemblages produced by a myriad of interactive causes (May, 1975). In this case, the fossil species represent a wide range of taxa from a wide range of environments. (Modified from Sepkoski, 1994.)

the estimates noted above (that at most a few per cent of species have been fossilized), the proportion of sampled species is even tinier. If 1% of total species have been fossilized, then the estimate is that 90% of sampled individuals come from just 0.0004% of the total number of species that have existed. Given that abundance promotes the likelihoods of both fossilization and paleontological sampling, this tiny fraction is strongly biased toward the most abundant species to have existed. While fossil abundance follows the log-series pattern familiar to ecologists (Koch, 1987; Carter and McKinney, 1992), even rare fossil species must have often been among the more abundant individuals in the original community.

On a positive note, it is worth emphasizing that, despite the apparent bias toward fossilization and discovery of abundant species, the fossil record contains many abundance patterns that are virtually identical to living species. It has been reviewed elsewhere (McKinney, in press) that the following abundance patterns of living species also appear as abundance fossil patterns:

- There is a 'hollow curve' of decreasing fossil abundance with increasing body size.
- Most fossil species are geographically restricted.

- Most fossil species have low abundance densities.
- Spatial and temporal fossil abundance variation patterns reflect living species variations (e.g. power law behaviour of fossil abundance time series and the 'species-area curve' that occurs when censusing fossil communities through time and space).
- There is a correlation of fossil geographical range with local abundance.

This implies that the biasing effects toward preservation of more common species does not eliminate useful information on the population dynamics of those fossil species that are often preserved and can be used to infer population and community processes (e.g. McKinney and Frederick, 1992; McKinney et al., in press).

7.3 FOSSIL SPECIES AND EXTINCTION SELECTIVITY

An ideal way to determine how rare species avoid extinction would be to estimate the longevity of various rare species in the geologic record and then identify traits that promoted greater longevity among certain rare species. Unfortunately, the strong bias against rare species renders both of these steps very difficult, and often impossible. Determining longevity requires knowing when (and where) rare species appear and disappear. But confidence intervals for first and last stratigraphic appearances of fossil species increase exponentially with decreasing abundance (Marshall, 1994).

Despite these problems, the fossil record can make a significant contribution to understanding how rare species avoid extinction. Paleontology has amassed a substantial literature on past extinction rates, and especially extinction selectivity, which can shed light on extinction in rare species. Much of the data is at coarse taxonomic and spatio-temporal scales. Beginning with Simpson's (1944) contributions to the grand evolutionary synthesis, and continuing through the many studies using Sepkoski's well known family and genus compilations (Sepkoski, 1992), much emphasis has been placed on comparative taxonomic extinction rates. This approach has been very successful at indicating taxonomic differences in extinction, such as the relatively high extinction rates of mammals, insects and ammonoids compared with foraminifera, bivalves and echinoids (Raup and Boyajian, 1988; Sepkoski, 1990; Stanley, 1990). Patterns of higher taxic turnover during mass extinctions are usually similar to those throughout most of geological time (Sepkoski, 1990; Norris, 1991). Of great immediate importance is that these different characteristic group extinction rates seem to carry over to current extinction rates (May et al., 1995). Mammals, birds and other taxa with high fossil extinction rates also have relatively high modern extinction rates (Chapter 8). This implies that past extinction taxonomic patterns may be of great relevance today.

7.3.1 Fossil extinctions and the phylogeny of rarity

Coarse extinction patterns are useful when asking questions of rarity extinction that are in a taxonomic or phylogenetic context. The rapidly growing use of the comparative method (Chapters 4, 13) has demonstrated the importance of accounting for phylogeny in asking such questions. A 'rare' insect species may contain many more individuals than a 'rare' mammal species, for instance. Similarly, is a 'rare' bivalve species less prone to extinction than a 'rare' mammal species, given that bivalves have a much longer average species duration (e.g. Stanley, 1990)? As a null hypothesis, the answer would be 'yes', although one would need to specify how rarity is being measured.

Similar questions can be asked at finer taxonomic scales, where patterns of selectivity are also emerging. Thus, extinction-prone species seem to be non-randomly clustered among genera, families and perhaps all levels of the taxonomic hierarchy (McKinney, 1995). For example, corals (Johnson *et al.*, 1995), foraminifera and brachiopods (Stanley and Yang, 1994) and echinoids (McKinney, 1995) are but a few of the clades that apparently have their more extinction-prone species non-randomly distributed among genera and families. Given that rare species may also be non-randomly distributed among genera and families (e.g. Lawton *et al.*, 1994), this may (at least partly) explain the extinction selectivity patterns seen in fossils at many taxonomic levels.

7.3.2 Resistance and resilience in rare species persistence

The large paleontological literature on extinction can also provide data about the underlying biological traits that produce taxonomic extinction selectivity. As Raup (1994) has noted, taxonomic selectivity 'carries the tacit assumption that genealogical relatedness implies similarity of physiology, ecology, or other attributes that determine susceptibility to extinction' (see also Sepkoski, 1990). Perhaps the most obvious example, just noted, is that rarity is such a phylogenetically inherited trait. But the focus of this paper is to dig deeper. Why do some rare species persist longer than others? In other words, what are other factors besides rarity that promote species survival? If one can empirically document traits that reduce extinction, and act independently of abundance, then these traits should, in theory, promote survivorship in those rare species that possess them.

Fortunately, the fossil record is rich with data about the role of biological traits besides rarity that increase extinction rates during both mass extinctions and 'normal' geological time. It is suggested that the rarity–survival traits during 'normal', or background, geological time are especially important because they tell us how extinction is avoided during the vast majority of time. Despite the popular interest in mass extinctions,

Raup (1994) estimates that 96% of all species extinctions occurred at times other than mass extinctions. In addition, many (perhaps most) of the traits that promote survival during background extinctions also promote survival during mass extinctions (e.g. Sepkoski, 1990; Norris, 1991). Jablonski (1995) has voiced similar views about the utility of background selectivity patterns in predicting species vulnerability during the current extinction crisis. How can we sort through the voluminous fossil data on extinction selectivity for 'rarity-independent' traits? The approach here is to compile the selectivity data and categorize them according to the different ways that they can promote survival in rare species. Specifically, Pimm (1991) discusses two key concepts, resistance and resilience, that are useful in describing how all species avoid extinction. As used here, **resistance** is the ability of a species to remain at its current abundance despite an environmental change. **Resilience** is the ability of a species to return to its former abundance after an environmental change. Neither resistance nor resilience are totally independent of abundance, but other traits besides abundance are also involved and can influence resistance and resilience. These include the 'rarity-independent' traits that can inform us how some rare species can delay extinction.

Figure 7.2 illustrates this. At a given abundance, including of course low abundance, species can vary in their resistance or resilience to environmental change. As extinction is the correlated deaths of individuals of a group within a relatively short period (Harrison and Quinn, 1989; Van Valen, 1994), traits that promote resistance or resilience can be seen as those that reduce the likelihood of correlated death of the entire group. As noted, abundance is one such trait because it increases the degree of correlation required to kill all individuals; but many other traits that promote resistance and resilience are at least partly independent of abundance. For example, Stanley (1990) notes that, besides rarity, increasing behavioural complexity promotes higher extinction rates. Similarly, Baumiller (1993) documents how coarse filter-feeding (at least in crinoids) can promote species survivorship. It is unlikely that the extinction-resistance promoted by such traits (and others discussed below) is entirely explained by their covariation with abundance.

An especially important trait that often promotes survival is being widespread. A species can be locally rare but occupy a wide area of extent (Gaston, 1994). Living in widely separated areas reduces the chance of a given number of individuals dying a correlated death from the same environmental disturbance. Another key trait is being locally rare but having individuals that are 'robust' or tolerant to environmental change. At a given level of rarity (however defined), the probability of correlated death decreases if each individual is less prone to death from relatively minor, and thus more frequent, environmental changes. Increased robustness can thus be measured as an increase in the magnitudes of environmental changes that an individual can survive, for a number of

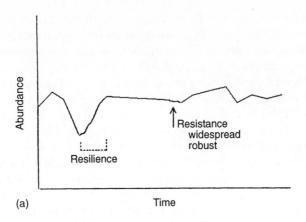

(a)   Time

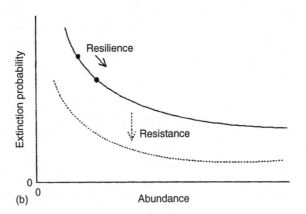

(b)   Abundance

Figure 7.2 (a) Resilience is measured by the time required to return to a former abundance. Resistance is measured here as the tendency not to show decreases in abundance. (b) Theoretical models and empirical data on extinction indicate an exponential decrease in extinction risk with increasing abundance (e.g. Leigh, 1981; Lande, 1993). Thus a resistant species can be viewed as one that is less likely to go extinct at a given level of abundance (rarity). On average, each individual in the resistant species has a lower probability of death than each individual in less resistant species, lowering the total probability of correlated death (Harrison and Quinn, 1989). Resilience is shown as the ability to return to higher abundances.

environmental parameters such as temperature, salinity and so on (Parsons, 1993).

Of course many of these 'rarity-independent' traits may themselves covary with each other. It is quite possible to possess a number of such traits that reduce extinction. A rare marine species could be a small detritivore, as well as occupy nearshore and productive habitat. All of these (as discussed below) could, in theory, synergistically interact to reduce greatly the chance of extinction in this rare species compared with a similarly rare species that lacked these traits.

7.4 RESISTANCE BY BEING LOCALLY RARE BUT WIDESPREAD

Geographical range and local population density are partially correlated (Chapters 3 and 4; Gaston, 1994). Many species that are locally rare therefore often have a restricted geographical range, but this is only a partial correlation and there are often many species in a group that are widespread but locally rare. In Gaston's (1994) terms, these species have a high extent of occurrence but a low area of occupancy. Such species are the focus of this section.

The advantage of being widespread in avoiding extinction is an idea at least as old as Lamarck, who noted that marine species may be less prone to extinction than land species because the former seem to be more widely dispersed. This advantage of being widespread has been generally supported by a growing body of ecological evidence (Chapter 8; Gaston, 1994). Most theoretical support has (at least implicitly) focused on the notion that being widespread reduces the chance of extinction from localized environmental changes. A recent supplement to this is Maurer and Nott's (in press) theory (and evidence) that rare species also suffer from relatively greater edge effects to their ranges. If true, being widespread, even if not densely populated, may be advantageous in reducing such edge effects.

Paleontological evidence on reduced extinction with increasing geographical range had accumulated by the early 1970s to the point where two books discussed fossil evidence for it. These were by Valentine (1973) and especially by Boucot (1975). Some of the later studies that have also shown this at the species level are listed in Table 7.1. Jablonski (1995) has a similar table of studies, but focusing on advantages of being widespread at the genus level. Probably the most widely cited fossil studies relating range to species longevity are Jablonski's work with marine molluscs. He showed that widespread species, with planktonic larvae, survive longer during most of geological time, but apparently not during mass extinctions (Jablonski 1986a,b, 1991). During mass extinctions, however, being widespread at the genus level does seem to promote survival. Westrop (1991) provides further examples of the pattern that being widespread at the genus and family level promotes survival, at least during most of geological time (outside of mass extinctions).

A potential sampling artefact of all the fossil data was noted by Russell and Lindberg (1988) who pointed out that, even if species have the same geological longevity, the species with the greater geographical range would

Table 7.1 Evidence that widespread species tend to have lower extinction rates in the geological record

Species	References
Mesozoic molluscs	Jablonski (1986a,b)
Late Mesozoic bivalves	Koch (1980)
Cenozoic bivalves	Jackson (1974)
Cenozoic gastropods	Hansen (1980)
Cenozoic foraminifera	Buzas and Culver (1991)

None of the data are limited solely to mass extinction selectivity; rather, they include survivorship during background times. Evidence that widespread range at the genus level promoted survival is listed in Jablonski (1995).

probably appear to have a longer duration due to sampling error. They performed simulations showing that, even with high levels of sampling, the more wide-ranging species would tend to have a longer stratigraphical range because it is more likely to be sampled by fossil collectors over a wider area. However, Marshall (1991) has re-examined the simulations of Russell and Lindberg, using a variety of analytical methods. Making what seem to be realistic assumptions about Jablonski's (1986a,b) data, Marshall finds that the correlation between geographical range and measured species longevity is apparently not solely artefactual.

As geographical range is correlated with local abundance, it is often difficult to prove, even in living species, that range alone is significantly enhancing species survivorship. Rosenzweig (1995) recounts only one study that he is aware of that explicitly seeks to disentangle these joint effects. The study (using ants on Barro Colorado Island, Panama) finds that range alone does enhance survival, independently of local abundance. In addition, the fossil record offers two other lines of evidence in support of the idea that geographical range contributes to survival of a species, independently of how locally abundant that species is.

One is by documenting the value of refugia in surviving extinctions. Vermeij (1993) discusses numerous examples of marine species that survived major habitat losses during late Cenozoic global climate changes by inhabiting geographical refuges. Such refuges usually occupied a small fraction of the former species range. Increased geographical range would thus increase the chances of occupying an area that would become a refuge during environmental change.

The second line of fossil evidence concerns 'Lazarus species', which are species that reappear in the fossil record after a long geological hiatus (Jablonski, 1986a). Rare species are the most likely to become Lazarus species, by being prone to sampling error. This indicates that rare species can survive for long geological periods. The most likely candidates for such Lazarus taxa are rare but widespread species because the migration of depositional areas indicates that very localized (geographically) species are not likely to be sampled and re-sampled in the fossil record (e.g. McKinney, in press; McKinney *et al.*, in press).

The common explanation for the evidence that geographical range promotes species survival is that a local disturbance cannot cause a species to become extinct if it is widespread. Up to two important additional factors come into play when widespread geographical range in a locally rare species is correlated with high dispersal (as with Jablonski's molluscan data) and/or highly successful local recruitment. With high dispersal or recruitment, extinction-resistance in a widespread locally rare species may be enhanced by increased gene flow and rapid recolonization after local extinction. A good example of the role of recruitment was recently noted in the paleontological literature where Edinger and Risk (1995) show that survivorship of coral genera over geologic time was substantially greater in corals with brooding larvae. Such corals apparently had significantly enhanced localized recruitment in stressful habitats.

7.5 RESISTANCE BY BEING ROBUST

While geographical range is evidently a good general predictor of extinction likelihood in rare species, there are clearly many other factors. Equally rare species, with exactly the same geographical ranges and densities, are unlikely to have exactly the same probability of extinction for a variety of reasons. Vermeij (1993) illustrates this with many recent anthropogenic extinctions where widely distributed species have become extinct, while more locally distributed species persist. The highly selective nature of human impacts (e.g. overhunting) may limit direct comparison with fossil extinctions, but it does demonstrate the existence of factors besides range. Discussions of vulnerability to extinction in the conservation biology literature specify many of these factors. These have at least some variance that is not entirely correlated to range patterns. Examples include trophic and life history traits, niche breadth, migratory patterns and many others (Chapter 8).

Robustness, in this chapter, refers to such non-range factors that contribute to the tolerance of individuals to environmental change. While amenable to direct measurement, such as tolerance to greater magnitudes of change in a number of environmental parameters (Futuyama and Moreno, 1988), robustness is most important here as an estimate of the probability of correlated death among many individuals. Specifically, which traits reduce the chance of correlated death from environmental change, when abundance and range are held constant? In terms of time-to-extinction models (e.g. Pimm *et al.*, 1993), being robust would increase time to extinction by reducing the frequency of environmental 'perturbations'. Leigh (1981), for example, shows that the more susceptible a population is to environmental fluctuations, the more slowly its expected lifetime increases with K (Lande, 1993 for review). This view is basically the same as Brown (1995) who sees increased niche breadth as a key factor that often promotes survival of individuals, populations and species.

Table 7.2 shows traits and habitats which, according to fossil evidence, seem to promote increased species duration at the relevant phylogenetic and temporal scale. This is an illustrative, not exhaustive, list culled from the extensive literature on extinction selectivity discussed above. Where possible, sources are listed that review the literature for the trait and habitat. This paleontological literature is the best direct source for data on robustness but is obviously limited to traits that are preserved. General discussions of trait and habitat selectivity in fossil background and mass extinctions are found in Jablonski (1991, 1995), Raup (1994) and Van Valen (1994). Many traits listed in the conservation biology literature as promoting extinction vulnerability (e.g. Meffe and Carroll, 1994; Chapter 8), such as behaviour and life history, are not visible in fossils. The traits and habitats in Table 7.2 are those for which there is substantial fossil evidence, and reflect selectivity at the appropriate phylogenetic and temporal scale for this chapter.

Perhaps not surprisingly, a general pattern of Table 7.2 is that eurytopy is the biotic trait commonly reported in the literature to promote survivorship in both aquatic and terrestrial taxa. As used here, **eurytopy** refers to being broadly adapted in a number of niche parameters including trophic breadth; or, in more common terms, a eurytope is a 'generalist'.

Table 7.2 Traits that apparently increase longevity of species by increasing robustness of individuals to environmental change

Biotas	Traits	References
Biotic traits		
Land and aquatic	Eurytopy	Stanley (1990); Eldredge (1992)
	Behavioural/morphological simplicity	Stanley (1973, 1990)
Aquatic	Detritivores	Van Valen (1994); Hansen *et al.* (1993)*
	Benthic larvae	Chatterton and Speyer (1989)
	Brooding larvae	Edinger and Risk (1995)
	Coarse-filter feeding	Baumiller (1993)
	Infaunal burrowing	Stanley (1986)
Habitat occupied		
High productivity		Vermeij (1986)
Nearshore marine		Jackson (1974); Aronson (1994)
Non-tropical		References in Ricklefs and Shluter (1993)

Unless noted, all these data include background extinction rates, indicating that 'on average' these traits promote longer species duration.
* = Data for mass extinction only.

Futuyama and Moreno (1988) and Brown (1995) discuss how such increased breadth often evolves to co-occur in many parameters simultaneously to produce species that are generalized in many ways, and are therefore robust to many environmental changes. It is an old notion, preceding Darwin, that specialization reduces the ability to adapt to environmental change. Specialization, or **stenotopy**, is a trait often noted in conservation biology as enhancing vulnerability. However, 'specialization' is a vague concept, often supported by anecdotal evidence in a tautological way. The fossil record is useful in providing direct quantitative evidence of biased extinction that may be related to breadth of adaptation. Eldredge (1992) and Stanley (1990) provide overviews of this fossil evidence. A widely discussed example is Vrba's (1987) documentation of biased extinction in stenotopic antelopes relative to a sister clade of eurytopic forms. Other examples of biased extinction in stenotopic species (compared with more eurytopic relatives) include corals (Kauffman and Fagerstrom, 1993; Edinger and Risk, 1995) and skates (Long, 1994). Given the evidence for 'Brown's hypothesis' (Lawton *et al.*, 1994; Brown, 1995) that eurytopic species are often more widespread and locally abundant than more specialized species, this explanation for extinction avoidance would seem to apply mainly to those rare species that are exceptions to the general correlation between abundance and niche-breadth. There is little data on just how many rare species are eurytopic (however measured), but we can infer that a rare eurytopic species probably has, on average, a significantly reduced chance of extinction compared with a rare stenotopic species.

Another general pattern of Table 7.2 is that morphological and behavioural simplicity seems to promote survivorship during 'normal' geological time (Stanley, 1973; 1990) and during mass extinctions (e.g. Norris, 1991; Arnold *et al.*, 1995). 'Simplicity' and the converse, 'complexity', are obviously very general traits and the data on such traits are for relatively coarse taxonomic scales. The evidence so far seems to imply that being relatively simple in terms of behaviour and/or morphology may be another way for a rare species to delay extinction.

Table 7.2 also shows a number of traits documented in aquatic organisms that may increase robustness. Compared with most other trophic categories, detritus-feeding organisms generally show low background extinction rates (McKinney, 1987; Van Valen, 1994) and a substantially higher survivorship during mass extinctions (Erwin, 1993; Hansen *et al.*, 1993). This may be attributed to a more generalized diet and lack of feeding specializations. That detritivores cope with stress better is indicated by their high relative abundance and taxonomic dominance during the two million years of environmental stress that followed the end-Cretaceous mass extinction boundary (Hansen *et al.*, 1993; but see Jablonski and Raup, 1995, for possible complicating factors). There is also (very tentative) evidence that being mobile promotes survivorship during background and mass extinction times (McKinney, 1987; Van Valen, 1994). Mobility has a partial correlation with detritivory, but many mobile aquatic invertebrates are not detritivores, such as grazers and carnivores.

In nearly all cases where trophic selectivity is studied (e.g. McKinney, 1987; Hansen *et al.*, 1993; Erwin, 1993), sessile suspension feeders have the highest extinction rate of any group during both background and mass extinctions. Within suspension feeders, Baumiller (1993) analysed a very large data base of Paleozoic crinoids that indicates a significantly higher background longevity for species that are coarse-filter feeders. He suggests that fine-filter feeders are more prone to extinction because they are more specialized in their diet and other traits. Other traits that could promote survival of rare aquatic species include certain reproductive traits such as brooding and/or benthic larvae, and infaunal burrowing (Table 7.2).

Finally, Table 7.2 shows three habitat categories that seem to experience higher background and perhaps mass extinction rates. The evidence for higher extinction rates of tropical habitats, both terrestrial and marine, is controversial for mass extinctions (Raup and Jablonski, 1993). Eldredge (1992) and Lawton *et al.* (1994) discuss a theoretical prediction as to why higher background tropical extinction rates may occur: tropical species are more endemic and stenotopic. Speculations about why high productivity promotes extinction resistance include the occurrence of a 'buffer' that would allow rare species to survive during times of stress (Vermeij, 1986). The higher resistance of nearshore species to extinction has been attributed to their being more eurytopic and widespread than offshore species (Jackson, 1974). Combining all this into a (simplistic) additive null model, we would predict that rare species that occupy highly productive, non-tropical, nearshore habitats would be most likely to persist.

In conclusion, all the above studies provide evidence for traits that promote 'robustness' by providing some resistance beyond that provided by any correlation they also have with increased geographical range. For example, extinction-resistance provided by detritivory, mobility, infaunal burrowing and coarse filter-feeding has usually been attributed to non-abundance factors so that all of these could aid the survival of rare species. It has thus often been suggested that detritivory and coarse filter-feeding allow individuals better to withstand environmental change because of a broader potential diet. Mobility and burrowing may promote survival by allowing individuals to escape physically from harmful environmental changes.

7.6 RESILIENCE BY BEING SMALL

Resilience, or the ability to regain individuals following a harmful environmental change, is strongly determined by age of maturation, growth rates and other life history variables (Pimm, 1991). Fossils rarely record most such variables, at least to the extent that extinction selectivity data have included them. A key exception is body size, which has drawn considerable paleontological attention. The purpose here is to review fossil evidence on size and extinction to see if it matches ecological predictions, and to show that small body size does seem to reduce extinction likelihood, during both normal and mass extinction intervals. Body size, like many traits noted above, is not totally independent of abundance (Gaston, 1994; Blackburn and Gaston, 1994). Nor is it clear that body size is completely independent of eurytopy and other traits that promote resistance. Because the fossil record so often omits life history data, it will probably be impossible to tease apart the role of each of these covariants in determining survival, especially in rare species.

We can, however, turn to ecological data for insight to explain fossil patterns. Blackburn and Gaston (1994) and Gaston (1994) discuss the complex influence of body size on extinction proneness. Various authors have asserted that larger body size can increase, decrease, or have no effect on probability of extinction. These conflicting assertions arise, in part, from comparing different temporal and spatial scales, just as they afflict paleontological selectivity studies. Once the various complexities are sorted out, it seems that large-bodied species suffer higher rates of historical extinctions (Blackburn and Gaston, 1994). The reason, apparently, is that the greater resilience provided by the shorter generation time of small body sizes tends to put species with large body sizes at greater extinction risk. The exception is when population size falls below seven pairs, at which point the greater resistance of larger body size outweighs its lower resilience (Pimm et al., 1988).

Ecological evidence therefore seems to imply that survival in rare species (or populations) will be promoted by smaller body sizes, as long as abundance exceeds a few pairs. The extreme biases of the fossil record against preservation and discovery of rare species makes it very unlikely that population extinction events involving just a few pairs (i.e. where large body size is favoured) will be visible in the record. We might thus expect that small body size will be recorded as favouring survival at most, and perhaps all, of the temporal and taxonomic scales visible in the record.

The fossil record does seem to generally support the extinction resistance of smaller body size at the coarse scales seen through geological time. Table 7.3 shows some of the studies that provide evidence for selectivity favouring smaller body sizes in many groups, at a variety of taxonomic and temporal scales. Additional studies that show this are cited in Jablonski and Raup (1995). Again, it is suggested (in agreement with Jablonski, 1995) that studies that focus on species level selectivity during background times are the most directly applicable to how rare species avoid extinction during the vast majority of time. This would seem to substantiate Stanley's (1979) claim that species longevity is generally inversely correlated with body size, even within clades. An exception is Van Valen's (1975) finding that genera of large mammals persist longer

Table 7.3 Evidence that large body promotes extinction relative to small body size at the coarse scales seen in geological time

Level	Group	References
Background extinctions		
Species	Late Cenozoic bivalves	Stanley (1986)
	Late Cenozoic mammals	Stanley (1979)
Genus	Foraminifera	Van Valen (1975)
*Exception:	Mammal genera	Van Valen (1975)
Mass extinctions		
Species	Late Mesozoic: planktic foraminifera	Norris (1991)
	Eocene–Oligocene: planktic foraminifera	Arnold et al. (1995)
	Pleistocene mammals	Martin (1984)
Clade	Late Paleozoic: tetrapods	Bakker (1977)
	Late Mesozoic: dinosaurs	Van Valen and Sloan (1977)
	Late Mesozoic: ammonites, rudist clams	Raup (1994)
*Exception:	Late Mesozoic: bivalve genera	Jablonski and Raup (1995)

Level refers to taxonomic level at which selectivity is reported. Thus, large-bodied species, genera or clades (such as dinosaurs) are observed to die out preferentially at mass extinctions, or have shorter durations during background time.

than smaller mammals (see also Martin, 1992 for similar but anecdotal results).

Note, however, that small body size seems generally advantageous during mass extinctions too. In this case, 'mass extinction' is not limited to the 'big five' that affected over 70% of all species in the biosphere (Jablonski, 1995). Rather, it refers to any catastrophic event that affected large numbers of species in the groups that are shown. The last Pleistocene, for example, is often seen as a 'mass extinction' when measured solely by its impact on megafauna (Martin, 1984). This broader definition is used here because such catastrophes provide data on how such groups reacted during stress, and thus which traits provide rare species with extinction resistance.

A notable exception to small body size advantage during mass extinctions is Jablonski and Raup's (1995) data on bivalve genera at the end-Cretaceous mass extinction. As these are genera and not species, it is not clear how this translates to survival at the species level, and this point is reinforced by the detailed study by Arnold et al. (1995). They find that planktonic foraminifera species appear to show random extinction with respect to body size during the Eocene–Oligocene transition, but closer inspection reveals that small-bodied species 'evolve' their way through the stressful period by speciating into new forms (Arnold et al., 1995). Arnold and others speculate that this preferentially occurs in small species because of their shorter generation times, whereas larger species become extinct. Given the often suggested ability of rare species, with smaller gene pools, to evolve relatively quickly (e.g. Stanley, 1990; Maurer and Nott, in press), such rapid evolutionary adaptation may be common. This conforms with fossil data showing that small mammals have higher speciation rates than large mammals, ostensibly because they have smaller geographical ranges (Martin, 1992).

7.7 SUMMARY AND DISCUSSION

The set of rare species may be biased by the selective elimination of species that cannot persist at low abundances. This chapter reviews the large paleontological literature on extinction selectivity for evidence that the set of rare species (however defined) is indeed biased in favour of those species with traits that promote species longevity by being resistant and/or resilient to disturbances.

These patterns indicate that, when rarity (abundance) is held constant, the following traits promote species survivorship relative to other species within the same clade.

- **Resistance** – ability of rare species to withstand disturbance.
  - ○ Wide geographical range – for species that are locally sparse.
  - ○ Robustness – traits that reduce susceptibility to environmental fluctuations, such as wide niche-breadth, morphological and behavioural simplicity, detritivory (and a suite of other traits in marine organisms).
- **Resilience** – ability of rare species to recover from disturbance. Fossils do not record many life history and other attributes of resilience, but body size data seem to support ecological theories that, above a very small population size, large-bodied species become extinct sooner.

A key question about these observations is the extent to which they are explained by the positive correlation of these traits with abundance. Small body size, wide geographical range and some robustness traits such as eurytopy are well documented as being correlated with increasing abundance (Gaston, 1994). Nevertheless, it seems unlikely that all of the survivorship patterns documented in fossil species are due to the abundance advantage associated with these traits. If so, then any rare species with these traits is more likely to persist than an equally rare species lacking them. Such traits explain why 'rarity alone is not a sufficient predictor of the probability of extinction of a species' (Gaston, 1994). Conversely, rarity can be a better predictor of extinction if information about such traits continues to be gathered and is added to models now based solely on abundance dynamics.

ACKNOWLEDGEMENTS

I thank Bill Kunin and Kevin Gaston for inviting me to participate in their very interesting symposium and book. I thank Bill Kunin, Julie Lockwood,

Tom Brooks, Gareth Russell and two anonymous referees for their comments on the manuscript. Funding from National Science Foundation Grant EAR-9316417 is gratefully acknowledged.

REFERENCES

Allison, P.A. and Briggs, D.E.G. (eds) (1991) *Taphonomy: Releasing the Data Locked in the Fossil Record*, Plenum, New York.

Arnold, A.J., Kelly, D.C., and Parker, W.C. (1995) Causality and Cope's rule: evidence from the planktonic foraminifera. *Journal of Paleontology*, **69**, 203–210.

Aronson, R.B. (1994) Scale-independent biological interactions in the marine environment. *Oceanography and Marine Biology: An Annual Review*, **32**, 435–460.

Bakker, R.T. (1977) Tetrapod mass extinctions: a model of the regulation of speciation rates and immigration by cycles of topographic diversity, in *Patterns of Evolution* (ed. A. Hallam), Elsevier, Amsterdam, pp. 439–468.

Baumiller, T.K. (1993) Survivorship analysis of Paleozoic Crinoidea: effect of filter morphology on evolutionary rates. *Paleobiology*, **19**, 304–321.

Blackburn, T.M. and Gaston, K.J. (1994) Animal body size distributions: patterns, mechanisms, and implications. *Trends in Ecology and Evolution*, **9**, 471–474.

Boucot, A.J. (1975) *Evolution and Extinction Rate Controls*, Elsevier, Amsterdam.

Brown, J.H. (1995) *Macroecology*, University of Chicago Press, Chicago.

Buzas, M.A. and Culver, S.J. (1991) Species diversity and dispersal of benthic foraminifera. *BioScience*, **41**, 483–489.

Carter, B.D. and McKinney, M.L. (1992) Eocene echinoids, the Suwannee Strait, and biogeographic taphonomy. *Paleobiology*, **18**, 299–325.

Chatterton, B.D. and Speyer, S.E. (1989) Larval ecology, life history strategies, and patterns of extinction and survivorship among Ordovician trilobites. *Paleobiology*, **15**, 118–132.

Edinger, E.N. and Risk, M.J. (1995) Preferential survivorship of brooding corals in a regional extinction. *Paleobiology*, **21**, 200–219.

Eldredge, N. (1992) Where the twain meet: causal intersections between the genealogical and ecological realms, in *Systematics, Ecology, and the Biodiversity Crisis* (ed. N. Eldredge), Columbia University Press, New York, pp. 1–14.

Erwin, D.H. (1993) *The Great Paleozoic Crisis*, Columbia University Press, New York.

Futuyama, D.J. and Moreno, G. (1988) The evolution of ecological specialization. *Annual Review of Ecology and Systematics*, **19**, 207–233.

Gaston, K.J. (1994) *Rarity*, Chapman & Hall, London.

Hansen, T.A. (1980) Influence of larval dispersal and geographic distribution on species longevities in neogastropods. *Paleobiology*, **6**, 193–207.

Hansen, T.A., Farrell, B.R. and Upshaw, B. (1993) The first 2 million years after the Cretaceous–Tertiary boundary in East Texas: rate and paleoecology of the molluscan recovery. *Paleobiology*, **19**, 251–265.

Harrison, S. and Quinn, J.F. (1989) Correlated environments and the persistence of metapopulations. *Oikos*, **56**, 293–298.

Jablonski, D. (1986a) Causes and consequences of mass extinctions: a comparative approach, in *Dynamics of Extinction* (ed. D.K. Elliot), Wiley, New York, pp. 183–229.

Jablonski, D. (1986b) Background and mass extinctions: the alternation of macroevolutionary regimes. *Science*, **231**, 129–133.

Jablonski, D. (1991) Extinctions: a paleontological perspective. *Science*, **253**, 754–757.

Jablonski, D. (1995) Extinctions in the fossil record, in *Extinction Rates* (eds J.H. Lawton and R.M. May), Oxford University Press, Oxford, pp. 25–44.

Jablonski, D. and Raup, D.M. (1995) Selectivity of end-Cretaceous bivalve extinctions. *Science*, **268**, 389–391.

Jackson, J.B.C. (1974) Biogeographic consequences of eurytopy and stenotopy among marine bivalve and their evolutionary consequences. *American Naturalist*, **108**, 541–560.

Johnson, K.G., Budd, A.F. and Stemann, T.A. (1995) Extinction selectivity and ecology of Neogene Caribbean reef-corals. *Paleobiology*, **21**, 52–73.

Kauffman, E.G. and Fagerstrom, J.A. (1993) The Phanerozoic evolution of reef diversity, in *Species Diversity in Ecological Communities* (eds R.E. Ricklefs and D. Schluter), University of Chicago Press, Chicago, pp. 315–329.

Koch, C.F. (1980) Bivalve species duration, areal extent and population size in a Cretaceous sea. *Paleobiology*, **6**, 184–192.

Koch, C. (1987) Prediction of sample size effects on the measured temporal and geographic distribution patterns of species. *Paleobiology*, **13**, 100–107.

Kunin, W.E. and Gaston, K.J. (1993) The biology of rarity: patterns, causes, and consequences. *Trends in Ecology and Evolution*, **8**, 298–301.

Lande, R. (1993) Risks of population extinction from demographic and environmental stochasticity and random catastrophes. *American Naturalist*, **142**, 911–927.

Lawton, J.H., Nee, S., Letcher, A. and Harvey, P. (1994) Animal distributions: patterns and processes, in *Large-scale Ecology and Conservation Biology* (eds P.J. Edwards, R.M. May and N. Webb), Blackwell, Oxford, pp. 41–58.

Leigh, E.G. (1981) The average lifetime of a population in a varying environment. *Journal of Theoretical Biology*, **90**, 213–239.

Long, D.J. (1994) Quaternary colonization or Paleogene persistence? Historical biogeography of skates in the Antarctic icthyofauna. *Paleobiology*, **20**, 215–228.

Marshall, C.R. (1991) Estimation of taxonomic ranges from the fossil record, in *Analytical Paleobiology* (eds N. Gilinsky and P. Signor), Paleontological Society, Knoxville, Tennessee, pp. 19–38.

Marshall, C.R. (1994) Confidence intervals on stratigraphic ranges: partial relaxation of the assumption of randomly distributed fossil horizons. *Paleobiology*, **20**, 459–469.

Martin, P.S. (1984) Prehistoric overkill: the global model, in *Quaternary Extinctions* (eds P.S. Martin and R. Klein) University of Arizona Press, Tucson, pp. 354–403.

Martin, R.A. (1992) Generic species richness and body mass in North American mammals: support for the inverse relationship of body size and speciation rate. *Historical Biology*, **6**, 73–90.

Maurer, B.A. and Nott, M.P. (in press) Geographic range fragmentation and the evolution of biological diversity, in *Biodiversity Dynamics: Turnover of*

Populations, Taxa, and Communities (ed. M.L. McKinney), Columbia University Press, New York.

May, R.M. (1975) Patterns of species abundance and diversity, in *Ecology and Evolution of Communities* (eds M.L. Cody and J.M. Diamond), Belknap Press, Cambridge, Massachussetts, pp. 81–120.

May, R.M. (1988) How many species are there on earth? *Science*, **241**, 1441–1449.

May, R.M., Lawton, J.H. and Stork, N.E. (1995) Assessing extinction rates, in *Extinction Rates* (eds J.H. Lawton and R.M. May), Oxford University Press, Oxford, pp. 1–24.

McKinney, M.L. (1987) Taxonomic selectivity and continuous variation in mass and background extinctions of marine taxa. *Nature*, **325**, 343–345.

McKinney, M.L. (1995) Extinction selectivity among lower taxa: gradational patterns and rarefaction error in extinction estimates. *Paleobiology*, **21**, 300–313.

McKinney, M.L. (in press) The biology of fossil abundance. *Revista Espanola de Paleontologia*.

McKinney, M.L. and Allmon, W.D. (1995) Metapopulations and disturbance: from patch dynamics to biodiversity dynamics, in *New Approaches to Speciation in the Fossil Record* (eds D. Erwin and R. Anstey), Columbia University Press, New York, pp. 123–183.

McKinney, M.L. and Frederick, D. (1992) Extinction and population dynamics. *Geology*, **20**, 343–346.

McKinney, M.L., Lockwood, J. and Frederick, D. (in press) Scale-dependence and rare species in community stasis. *Palaeogeography, Palaeoclimatology, and Palaeoceanography*.

Meffe, G.K. and Carroll, C.R. (1994) *Principles of Conservation Biology*. Sinauer, Sunderland, Massachusetts.

Norris, R.D. (1991) Biased extinction and evolutionary trends. *Paleobiology*, **17**, 388–400.

Parsons, P.A. (1993) Stress, extinctions, and evolutionary change: From living organisms to fossils. *Biological Reviews*, **68**, 313–333.

Pimm, S.L. (1991) *The Balance of Nature?*, University of Chicago Press, Chicago.

Pimm, S.L., Jones, H.L. and Diamond, J. (1988) On the risk of extinction. *American Naturalist*, **132**, 757–187.

Pimm, S.L., Diamond, J., Reed, T. *et al.* (1993) Times to extinction for small populations of large birds. *Proceedings of the National Academy of Science*, **90**, 10871–10875.

Raup, D.M. (1988) Diversity crises in the geological past, in *Biodiversity* (eds E.O. Wilson and F.M. Peter), National Academy Press, Washington, DC, pp. 51–57.

Raup, D.M. (1991) *Extinction: Bad Genes or Bad Luck?*, W.W. Norton, New York.

Raup, D.M. (1994) The role of extinction in evolution. *Proceedings of the National Academy of Science*, **91**, 6758–6763.

Raup, D.M. and Boyajian, G. (1988) Patterns of generic extinction in the fossil record. *Paleobiology*, **14**, 109–125.

Raup, D.M. and Jablonski, D. (1993) Geography of end-Cretaceous bivalve extinctions. *Science*, **260**, 971–973.

Ricklefs, R.E. and Schluter, D. (1993) Species diversity: regional and historical influences, in *Species Diversity in Ecological Communities* (Ricklefs, R.E. and Schluter, D.), University of Chicago Press, Chicago, pp. 350–364.

Rosenzweig, M.L. (1995) *Species Diversity in Space and Time*, Cambridge University Press, Cambridge.

Russell, M.P. and Lindberg, D.R. (1988) Real and random patterns associated with molluscan spatial and temporal distributions. *Paleobiology*, **14**, 322–330.

Sepkoski, J.J., Jr (1990) The taxonomic structure of periodic extinction, in *Global Catastrophes in Earth History* (eds V.L. Sharpton and P.D. Ward), Geological Society of America, Boulder, Colorado, pp. 33–44.

Sepkoski, J.J., Jr (1992) *A Compendium of Fossil Marine Animal Families*, Milwaukee Public Museum Contributions to Biology and Geology, Volume 83.

Sepkoski, J.J., Jr (1994) Limits to randomness in paleobiologic models the case of Phanerozoic species diversity. *Acta Palaeontologica Polonica*, **38**, 175–198.

Simpson, G.G. (1944) *Tempo and Mode in Evolution*, Columbia University Press, New York.

Stanley, S.M. (1973) An explanation for Cope's Rule. *Evolution*, **27**, 1–26.

Stanley, S.M. (1979) *Macroevolution*, W.H. Freeman, New York.

Stanley, S.M. (1990) The general correlation between rate of speciation and rate of extinction: fortuitous causal linkages, in *Causes of Evolution* (eds R. Ross and W.D. Allmon), University of Chicago Press, Chicago, pp. 103–127.

Stanley, S.M. and Yang, X. (1994) A double mass extinction at the end of the Paleozoic Era. *Science*, **266**, 1340–1344.

Valentine, J.W. (1973) *Evolutionary Paleoecology of the Marine Biosphere*, Prentice-Hall, Princeton, New Jersey.

Van Valen, L.M. (1975) Group selection, sex, and fossil. *Evolution*, **29**, 87–94.

Van Valen, L.M. (1994) Concepts and the nature of natural selection by extinction: Is generalization possible?, in *The Mass Extinction Debates* (ed. W. Glen), Stanford University Press, Stanford, pp. 200–216.

Van Valen, L.M. and Sloan, R.E. (1977) Ecology and extinction of the dinosaurs. *Evolutionary Theory*, **2**, 37–64.

Vermeij, G.J. (1986) Survival during biotic crises: the properties and evolutionary significance of refuges, in *Dynamics of Extinction* (eds D.K. Elliott), Wiley, New York, pp. 231–246.

Vermeij, G.J. (1993) Biogeography of recently extinct marine species: implications for conservation. *Conservation Biology*, **7**, 391–397.

Vrba, E.S. (1987) Ecology in relation to speciation rates: some case histories of Miocene – Recent mammal clades. *Evolutionary Ecology*, **1**, 283–300.

Westrop, S.R. (1991) Intercontinental variation in mass extinction patterns: influence of biogeographic structure. *Paleobiology*, **17**, 363–368.

8 Extinction risk and rarity on an ecological timescale

Georgina M. Mace and Melanie Kershaw

8.1 INTRODUCTION

The concept of rarity continues to have many different applications in the biological literature. It has long been recognized that rarity can be measured on several different axes, such as ecological specialization and distribution (Mayr, 1963), or local population size, habitat specificity and geographical distribution (Rabinowitz *et al.*, 1986), as well as others less commonly identified (Chapter 3; Gaston, 1994). In addition, because these are all features that vary continuously, rarity should also be regarded as a continuous variable, and Gaston (1994) has recently provided some practical proposals for its definition and analysis. The purpose of this chapter is to review the way in which rarity is associated with persistence, and how evolutionary pressures have shaped the kinds of rarity we see represented in extant species.

Although it is widely assumed that rarity is associated with increased extinction risk, this is not necessarily the case. Populations may be small in abundance or range because they are censused at a time and place where they are in decline from previously high numbers, or because they are at the edge of their range and are therefore occupying sub-optimal habitats. We assume here that, apart from extrinsic processes that may eliminate species whatever their biological characteristics, some kinds of rarity do allow populations to persist over long periods. Rare species that have persisted are expected to exhibit characteristics that allowed them to become rare, and then enabled them to persist in a rare state (Kunin and Gaston, 1993). The problem with an exhaustive study of the life history and ecological characteristics that allow species to pass through these filters is that there are many ways in which a species might be considered 'rare' and making a single cut may constrain the results of the analysis. We take a different approach and ask which kinds of 'rare' species are likely to be lost through local or global extinction, and therefore which kinds of rarity are

The Biology of Rarity. Edited by William E. Kunin and Kevin J. Gaston. Published in 1997 by Chapman & Hall, London. ISBN 0 412 63380 9.

likely to persist to be observed and recorded. We begin by reviewing the studies that have identified correlates of persistence in cross-species studies. We then examine ecological characteristics and conservation status in a set of South African birds to identify rarity correlates of high extinction risk. Finally, using the global list of threatened bird species, we examine the rarity correlates of species that are currently at high risk of extinction, or which are predicted to decline rapidly in numbers or range in the future. The set of attributes that characterize 'rare' species at risk of extinction might therefore be those that are not characteristic of species that persist.

From basic theoretical models of population persistence, a relationship between components of rarity and persistence is expected. Birth/death models predict that population lifetimes will increase with generation length, population size, the intrinsic rate of increase (r) and low variation in r (Richter-Dyn and Goel, 1972; Leigh, 1981; Goodman, 1987), and several studies have examined these associations empirically. Generally this has been done by comparing the characteristics of species that persist or are secure in some habitat that has undergone fragmentation or isolation, with those that are missing from it or are threatened within it. These studies have shown that many different factors are associated with persistence, including those factors listed above that are expected from basic theory, and other variables closely associated with these. For example, studies have shown that species are more vulnerable to local extinction when they have restricted ranges or occupy small numbers of sites (Simberloff and Gotelli, 1984; Thomas and Mallorie, 1985; Digby and Kempton, 1987; Happel et al., 1987; Thomas, 1991), are endemic to an area or show particular ecological specialization (Terborgh and Winter, 1980; Cowling and Bond, 1991; Laurance, 1991), have low abundances (Diamond, 1984; Pimm et al., 1988; Bolger et al., 1991; Given and Norton, 1993) and have poor dispersal ability and high population variability (Karr, 1982; Diamond, 1984).

There are several problems in using these studies for the relationship between rarity and persistence. The first is that the different variables associated with persistence are strongly interrelated. Thus, ecological specialists tend to also have low abundances, small ranges, low dispersal ability and, often, small body size and short lifespans (Diamond, 1984; Laurance, 1991; Gaston, 1994), and it has proved to be hard to disentangle them, as well as to distinguish between the causes as opposed to the consequences of rarity. Secondly, almost all such studies are performed at local scales, and may represent habitats where species regularly undergo local extinction and colonization, but where a dynamic equilibrium is maintained (Diamond, 1984). This is particularly likely for continental island archipelagos. Some studies have focused on habitats that have recently undergone extreme alteration, resulting in the local extinction of species with particular ecological specializations. In these cases the characteristics of the vulnerable species may not necessarily be those that are generally associated with increased extinction risks. Finally, as Gaston (1994) notes, most of these data sets were compiled from general censuses that did not control for differences between species in habit, habitat, abundance or distribution, any of which might contribute to their being either under- or over-represented in the sample. It is therefore possible that some associations found may be statistical artefacts, or that the results may alter depending on the scale of the area under study (Saetersdal, 1994).

Complex as it may be in the ecological literature (Chapter 3; Kunin and Gaston, 1993), the role of rarity in the recognition and classification of threatened and endangered species is even more confused. To some, rarity and threat are almost synonymous, while others recognize that rarity *per se* is insufficient to classify a species at high extinction risk, and that the recognition of threatening processes causing continuing population decline is the critical component (Munton, 1987). However, a significant practical application for an understanding of the causes and consequences of rarity is in the setting of conservation priorities based upon an assessment of extinction risk. We therefore review below the way in which rarity has been and is currently incorporated into threatened species listing. We then go on to present some analyses on the relationship between various components of rarity and qualitatively or quantitatively based assessments of extinction risk.

8.2 RARITY AND THREAT DEFINITIONS

In this chapter, we analyse the rarity correlates associated with global extinction risk in bird species. To avoid circularity in this analysis it is important to understand the way in which correlates of rarity, such as small population size, ecological specialization and small range size, contribute to a species being listed as 'threatened'. In fact, lists of threatened species, such as the IUCN Red List (IUCN, 1994a) and its predecessors, have commonly included a 'Rare' category (Munton, 1987). 'Rare' has always been distinct from the 'Endangered' and 'Vulnerable' categories that are explicitly defined as species 'in danger of extinction', and is for species threatened because of their 'small world populations' (IUCN, 1994a). Although the definition is based on population sizes, the notes accompanying the definition stress that the category is intended for species that are 'localized within restricted geographical areas or habitats or thinly scattered over a more extensive range' (IUCN, 1994a). Unlike the other threat categories there is no presumption that the species is at risk from identified threats and/or processes. Rather species are presumed to be at increased extinction risk simply because of their rarity. This is a questionable assumption. Some classically rare species have persisted over long periods, and systematic studies have shown that the dominant contributors to extinction risk are extrinsic processes driving population declines and not the population size itself (Chapter 11; Gilpin and Soulé 1986; WCMC, 1992; Given and Norton, 1993; Caughley, 1994). In addition, rare species have different characteristics from common ones that may represent adaptations to rarity (Rabinowitz, 1981; Munton, 1987; Rabinowitz et al., 1989), although it is also possible that these are causes rather than consequences of rarity (Kunin and Gaston, 1993) and therefore do not represent adaptations.

The IUCN threatened species categories have recently been revised, and a different approach taken over the treatment of rare species. The revised categories are focused only on the assessment of extinction risk (Mace and Lande, 1991). Categories such as 'Rare', which describe an aspect of population status rather than an assessment of extinction risk, no longer exist. Instead, the particular characteristics that contribute to extinction risk, such as small population size or restricted distributions, are factored into quantitative criteria presented for each category of threat. Under the revised system (IUCN, 1994b), it is unlikely that rare species qualify as threatened unless there are also observations of continuing decline or reasons to expect future declines. Only species with very small total population sizes (fewer than 1000 mature individuals) can be listed as threatened when there is no likelihood of population decline. Species with very restricted distributions can be listed at only the lowest threat level (Vulnerable). There are good reasons for these criteria. Clearly, it is almost a contradiction to list species as at risk of extinction when there is no evidence for any existing and/or future declines in abundance or range. Once any such evidence does emerge, species with low abundance or range size will immediately qualify, often at a high category of threat (Collar et al., 1994). However, there is a point at which extinction risks do increase in small populations from entirely intrinsic processes, such as the effects of demographic stochasticity, loss of genetic variation and social dysfunction (Soulé, 1987; Lande, 1988). On this basis there are criteria for listing species in threatened categories because of factors related to natural rarity, though these criteria are set at quite low threshold levels (IUCN, 1994b).

One potentially confounding factor that deserves further investigation is the extent to which populations that have always been rare may be better adapted to rarity, and therefore at lower extinction risk, than those that have been recently reduced from previously wide-ranging or highly abundant populations (Lawton, 1995). For example, many Australian marsupials have declined dramatically from historical levels and are now restricted to a small fragment of their former range (Burbidge and McKenzie, 1989; Kennedy, 1992). However, because they are still quite numerous, and the historical declines are many generations in the past, they no longer qualify as threatened according to the new IUCN criteria. All other things being equal, species that have been substantially reduced might warrant higher threatened status than those that have never been abundant, and the importance of past events in shaping species' ability to

persist under subsequent environmental changes is an important area for
further study (Balmford, 1996).

Useful analyses are compromised by the multiple usages of the term
'rarity'. For practical purposes it would be better to define what aspect of
rarity (population size, range size or habitat type) is important in the
particular case, and what quantitative threshold is being used to denote
rarity (Gaston, 1994). The next section investigates which of several
measures of rarity are most closely associated with threatened status
among passerine birds in KwaZulu Natal. Here the assessment of
threatened status is not explicit and might be based on any of many factors
considered influential by the local biologists making the assessment.
Section 8.4 investigates the extent to which different kinds of rarity
correlate with estimates of high or low extinction risk. This analysis is
based upon data on all threatened bird species formulated from the new,
quantitative IUCN threatened species criteria (Collar *et al.*, 1994; IUCN,
1994b).

8.3 ANALYSIS OF RARITY AND THREAT AMONG PASSERINE BIRDS IN KWAZULU NATAL

8.3.1 Introduction

The purpose of this analysis is to investigate the extent to which different
kinds of rarity contribute to the listing of species as threatened, as a
surrogate for extinction risk. Unfortunately, it is hard to avoid some
circularity as threatened status is decided either explicitly or implicitly from
some population and/or distributional state. In the extreme case this
analysis might simply (and trivially) reveal the rules used. To reduce this
possibility the study focused on an area and fauna where all rarity variables
could be measured directly and independently of the threat classification,
and where the threat classification used qualitative rules and consistent
analyses by field biologists for species listing.

Ideally, such an analysis would be undertaken on a wider scale than just
within one country. However, ecological variables related to local
abundance and habitat specificity are similar across species' ranges, and
therefore their assessment within one country can be used to score them
more generally. The distribution of range sizes within Natal is assumed to
be representative of their distribution on a global basis, at least in relative
terms.

8.3.2 Methods

The analysis was based on passerine birds in KwaZulu Natal, South Africa.
KwaZulu Natal is located on the east coast of South Africa and has a total
area of approximately 91 800 km². It has a well studied and documented

fauna and flora, and high levels of diversity and endemism. The 630 species of birds that have been recorded represent 70% of the total for southern Africa in just 2.6% of the area (Sinclair *et al.*, 1993); 54% of the birds endemic to southern Africa occur in KwaZulu Natal.

Passerine birds were selected for this analysis as they form a relatively homogeneous group, in both their biological characteristics and the data that are available for them. Species distribution data were extracted from a bird atlas based upon 10 years of fieldwork (Cyrus *et al.*, 1980). All vagrants and introduced species were excluded as were marine or aquatic forms. This left a total of 195 species. Each of these was scored for the following attributes.

- **Threatened status**. All species listed as globally threatened (Collar and Stuart, 1985; IUCN, 1994a) were scored as threatened, including those listed as near-threatened or indeterminate. Others were scored as not threatened. Species only listed as threatened within the regional South African lists (Brooke, 1984) were not included.
- **Endemism**. Species with ranges entirely restricted to southern Africa were scored as endemic. This list includes species with distributions south of the River Kunene and River Zambezi, in Namibia, Botswana, Zimbabwe, South Africa, Lesotho, Swaziland and central and southern Mozambique (Sinclair *et al.*, 1993). A few species with ranges that extend slightly outside this region – mostly into southern Angola – were also scored as endemic.
- **Abundance**. This measure referred to the local population size, and species were scored as either Common, Locally Common or Scarce, based on descriptions in the atlas and field guides. The categories are similar to the measures used by Rabinowitz *et al.* (1986), who used 'somewhere large' versus 'everywhere small'. Here, we distinguish between 'common' species that are widespread and always abundant, 'locally common' species that are found less frequently, but with large population size when found, and 'scarce' species that are always encountered infrequently.
- **Range**. Distribution data on species within KwaZulu Natal were taken from a bird atlas (Cyrus *et al.*, 1980) which records the presence of each species in each quarter degree grid square. This is an area of 24 × 27.5 km in South Africa, and so gives squares with an area of 660 km^2. KwaZulu Natal contains 166 of these squares. The number of squares occupied was used as a score of distribution size for each species. Besides this continuous measure, species were classified into four quartiles of the distribution. The lowest quartile was then used as a restricted range measure (Gaston, 1994). Figure 8.1 shows the frequency distribution of range sizes. The 25% quartile boundary was at 23 squares, so all species with ranges restricted to 23 or fewer grid squares were scored as restricted.

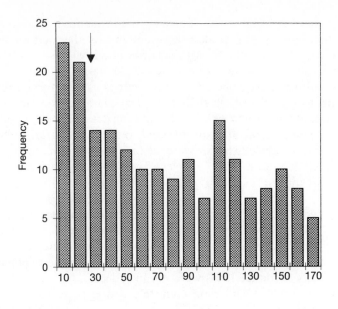

Number of grid squares

Figure 8.1 Frequency distribution of number of ¼ degree grid squares occupied by 199 passerine bird species in KwaZulu Natal, South Africa. Arrow shows lowest quartile (less than 21).

- **Habitats**. Using the field guides and atlases (Cyrus *et al.*, 1980; Sinclair *et al.*, 1993), species were scored for the number of habitats they were recorded in. These habitats were broadly defined into classes such as forest, upland grassland, montane, woodland and thornveld. Species were classified as habitat specialists if they occupied only one habitat type or habitat generalists if they occupied more than one.

These variables were analysed using a generalized linear model to predict the threatened status (threatened versus not threatened) from the four measures of rarity (Figure 8.2). Logistic regression was used since threatened status is a binary variable (McCullagh and Nelder, 1989). A full model was fitted that accounted for the maximum amount of deviance, and the independent variables were then dropped one by one to estimate the contribution of each to total deviance. The difference in the deviance of two models without the variable in question is distributed as chi squared (χ^2) with degrees of freedom equivalent to those of the variables being tested (Crawley, 1993).

8.3.3 Results

Table 8.1 shows the threat and rarity classifications for the 195 species of passerine bird in KwaZulu Natal. There were relatively few threatened

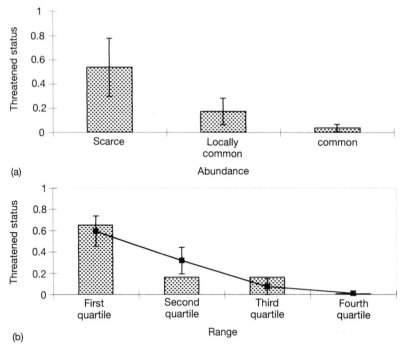

Figure 8.2 Predictions from a general linear model explaining variation in threatened status from variation in abundance, range size, endemism and habitat specialization. (a) Predictions about the probability of a species being threatened on the basis of a ranking of population abundance types showing the predicted score and the standard error. Range was set at the median value of 59, and species were assumed to be endemic. (b) Predictions about the probability of a species being threatened on the basis of range size. This histogram represents predicted values when range was scored as four ranks representing each quartile of the distribution. The lines represent estimated values and standard errors for a linear model where range size is a continuous variable. Abundance was set at 'Locally Common' and species were assumed to be endemic.

Table 8.1 One-way classifications of 195 passerine bird species in KwaZulu Natal

Variable	Value	Frequency
Threatened status	Threatened	14
	Not threatened	181
Endemism	Endemic	46
	Not endemic	149
Abundance	Common	138
	Locally common	41
	Scarce	16
Habit	Generalist	127
	Specialist	68
Range	Restricted	48
	Widespread	147

species (14) and in each other variable the 'rare' category was, as expected, under-represented. This meant that there were only small sample sizes in the analysis for some combinations of rarity and threat.

Some variables in the model were very closely associated (Table 8.2). In particular Abundance (measuring local abundance) and Range (measuring the distribution within Natal) were closely associated so that 'common' species are unlikely to have limited range sizes. 'Common' species were also most likely to be more generalized in habitats, whereas 'locally common' and 'scarce' species were more likely to be specialized. Endemic species were more likely to be specialized in habitat types.

The generalized linear model includes all these variables as predictors of threatened status, and several different models were fitted to the data to find the one that was most efficient at explaining threatened status by maximizing the total deviance accounted for. In this best model, range was fitted as a continuous variable (Table 8.3), and range, endemism and abundance all contributed significantly to threatened status, but habitats did not. Among the significant variables, threatened status was associated most strongly with endemism and abundance, and only weakly with range.

Table 8.2 Associations between ecological variables used in the threat analysis for birds in KwaZulu Natal, with chi-squared values, degrees of freedom and probability levels for each pair of variables (range size taken as a two-value score for each species based on whether its range size lies in lowest quartile of continuous distribution or upper three quartiles)

	Endemism	Habit	Range
Abundance	1.01, 2 df	57.4, 2 df	32.83, 2 df
	ns	$P < 0.001$	$P < 0.001$
Endemism	–	4.45, 1 df	2.07, 1 df
		$P < 0.05$	ns
Habitats	–	–	0.62, 1 df
			ns

Table 8.3 Analysis of deviance table for threatened status (range as a continuous variable with species represented in between one and 166 grid squares)

Variable	Deviance (χ^2)	df	P
Habitats	0.77	1	ns
Range	3.58	1	0.058
Endemism	15.86	1	< 0.001
Abundance	12.99	2	< 0.001

Model $\chi^2 = 36.179$, 2 df, $P < 0.001$

8.3.4 Discussion

As might be anticipated, all the rarity variables in this analysis were closely associated across species. Perhaps most surprising here was the lack of a significant association between endemism and range. This results partly from the fact that range size was measured on only a subset of the total range of most species in the analysis, although we would expect this measure to be correlated with their total range sizes. The independence of these two variables also reinforces the point that while endemism measures restriction in the total range of a species and therefore sets a limit on range size, widespread species may comprise many small populations distributed over a wide area. Although the total range size may be large, the actual area (in this case, grid squares) occupied could be quite small. Therefore, endemic species are likely to have small range sizes, but small range size species are not necessarily endemic.

When the variables are entered into a multivariate analysis a strong association is found between endemism and threatened status, and between population abundance and threatened status. The contribution of range size is weak and habitat specialization is not significant. The habitats measure we used here provides only a coarse indicator of local habitat usage, and therefore of habitat specificity. From the results presented here, we could conclude that the best predictor of threatened status is endemism and population abundance.

There are some important caveats to this analysis. First, the focus here is on passerine bird species in KwaZulu Natal, and previous studies have shown that different relationships may exist within different taxa or at different spatial scales (Saetersdal, 1994). The generality of these conclusions remain to be tested. Second, the measures of habitat specialization especially were quite coarse, and might not have been sufficiently detailed to identify genuine ecological specialists. Finally, we have to assume that the assessments of threatened status were made without any direct use of the kinds of measures used here. For example, if population abundance, as scored here, had been consistently used in the assessment of threat levels then a proven association would be a trivial finding. From the published accounts, we find no reason to suspect that there was any consistent bias of this kind for any of the variables examined, with the exception of endemism. It has been common, and understandable practice, for regional Red List compilers to include endemic species in their threatened species lists preferentially (Mace, 1994), without there being any other reason to suspect threatened status. The importance of endemism for threatened status may be overstated in this analysis.

Therefore, if these threatened categories do provide a reliable assessment of future extinctions, then the class of rare passerine bird species that will be lost at the highest rate will be those with small total or local population sizes. The implications of this are that, of the possible kinds of rarity, the class most likely to be heading for extinction is one of those with limited populations rather than those with limited total ranges, localized distributions (endemism) or specialized habitats. Under these circumstances, of the species that become rare, extinction biases would favour the persistence of those with local or specialized distributions.

8.4 ANALYSIS OF SEVERITY OF THREAT AMONG BIRDS

8.4.1 Introduction

To examine the severity of threat among rare species and the kinds of rarity that are most vulnerable to extinction pressures, we need to have some way of comparing extinction risk across species that exhibit different kinds of rarity. One method for assessing the likelihood of extinction would be a quantitative analysis incorporating aspects of the life history and ecology of a species, the threats it faces and the interactions between these. However, such specific analyses are complicated to undertake and require a substantial amount of information (Soulé, 1987). For global assessments, more general rules are used to classify species into categories of extinction risk.

The recently reformulated IUCN threatened species categories (IUCN, 1994b) provide quantitative criteria for three categories of extinction risk (Critical, Endangered and Vulnerable) in terms of population decline rates, population sizes and distribution (Table 8.4). These criteria can be used broadly to represent different kinds of rarity, and are used here to examine whether more highly threatened species are disproportionately

Table 8.4 Outline of the new IUCN Categories of Threat

Any of		Threatened classification
Criterion	Population status	
A1	Rapid decline observed or inferred	
A2	Rapid decline predicted or anticipated	
B	Small distribution with decline, fragmentation and fluctuation	Critical (CR)
C	Small population size with decline	Endengered (EN)
D1	Very small population	Vulnerable (VU)
D2	Very restricted distribution	
E	Unfavourable population viability analysis	

Any of a number and combination of states and processes can lead to a species being listed in one of the threatened categories of Critical, Endangered or Vulnerable. Each criterion (A–E) has numerical thresholds for each of the categories, which can be found in IUCN (1994b). The criteria (A to E) as well as the categories (CR, EN and VU) are used in this analysis.

composed of species with restricted population sizes, restricted geo-graphical ranges or both of these. To qualify as threatened, species need to meet any one of these criteria, and not meeting any other has no bearing on the resulting threat category. It is recognized that some criteria are more relevant for particular kinds of species and that it will frequently be impossible to assess some species by criteria for which data are simply unavailable (Mace, 1994). However, for birds, it was possible to assess most species by most of the criteria because of the exceptional database on birds that led to the compilation of Collar *et al.* (1994). Here, therefore, it is justifiable to analyse the criteria data on the basis that non-listing of a criterion can indicate that the criterion was not met, rather than that the data were not available to assess the species against this criterion. This is unlikely to be possible for other major taxa until our knowledge about most species increases substantially.

8.4.2 Methods

Species data were compiled from a recent conservation review of birds (Collar *et al.*, 1994) which includes 1111 species categorized as threatened according to the new IUCN criteria (IUCN, 1994b). For each species this compilation includes the category of threat (Critical, Endangered or Vulnerable), and the criteria (A, B, C, D or E; see Table 8.4) by which the species qualifies as threatened (i.e. is at least Vulnerable) as well as criteria by which it qualifies at a higher threat level (Critical or Endangered). The criteria for Vulnerable listing formed the basis of this analysis.

To examine how threatened status varies with restricted population size, restricted geographical ranges, or both or neither of these, we compared the proportion of species in each threat category (Vulnerable, Endangered or Critical) that qualified by the different criteria.

Secondly, a different approach was taken using different listings of the criterion A (Table 8.4). This criterion is based upon extreme population reduction without any limit upon the size of the population. Any species that is 'observed, estimated, inferred or suspected to have declined by at least 20% over the last 10 years or three generations (whichever is the longer)' is listed as Vulnerable by criterion A1, and any species that is 'expected' or 'projected' to decline at least at this rate is listed as Vulnerable under criterion A2. For A1 or A2 listing, the assessment can be based on any of a variety of measures from direct observation (A1 only), indices of abundance, levels of exploitation or decline in the area, extent or quality of habitat (for details see IUCN, 1994b; Collar *et al.*, 1994). Therefore criteria A1 and A2 provide a judgement of whether a species has been in rapid decline and is expected to continue to decline at this rate (A1 and A2), has been in rapid decline that is now slowed or halted (A1 only), is now about to enter a period of high risk of rapid decline that was not observed before (A2 only), or is not now nor has been recently in rapid decline (neither A1 nor A2). On the basis that species listed under A2 are the ones now likely to be in rapid decline, the extent to which A2 species were also listed because they qualified under the small population criteria (C and D1) or the restricted range criteria (B and D2) were contrasted to examine rarity correlates of rapid current declines.

8.4.3 Results

A total of 1107 species was classified into the threatened categories of Critical, Endangered and Vulnerable by Collar *et al.* (1994); four species were classified as Extinct in the Wild. Table 8.5 shows the frequency with which species were classified as threatened by each criterion. The most common criterion for birds is the population size criterion C. This is also the criterion that most often leads to threatened status when a species qualifies by one criterion only.

The population size criteria C and D1 remain the most frequently cited when the degree of threat is taken into account, although their relative frequencies vary with the threat category (Figure 8.3). Criterion D1, which is triggered in species with total populations of less than 1000 mature individuals with no evidence for continuing decline, was most commonly cited for species in the most threatened category, Critical. However, the frequency of D1 declined rapidly in the lower threat categories (Endangered and Vulnerable). In contrast, the criterion C was met most commonly by species qualifying in the lower threat categories of Endangered and Vulnerable. To qualify as threatened by criterion C, species need to be estimated at fewer than 10 000 mature individuals, and there needs to be evidence of continuing decline and/or fragmentation though there is no threshold for the rate of decline and it may be slight. The criteria reflecting limited range sizes (B and D2) were met less frequently at all levels of threat (Figure 8.3).

Figure 8.4 shows the frequency with which species that meet criterion A2 (expected rapid future decline) also meet criteria B, C, D1 or D2. The first histogram (criterion total) shows that across all taxa, A2 was met most

Table 8.5 Frequency with which different criteria (A to D) were used in classification of 1107 bird species as Critical, Endangered or Vulnerable by Collar *et al.* (1994)

Criterion	Number of species qualifying as threatened by this criterion	Number of species by this criterion only
A1	319	23
A2	285	13
B	422	45
C	764	201
D1	351	61
D2	256	86

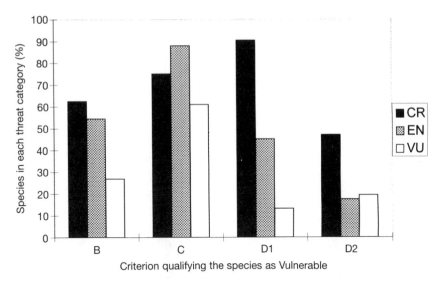

Figure 8.3 Percentage of threatened birds that qualify for different levels of threat (CR, EN or VU) by meeting alternative quantitative criteria according to new guidelines for categorizing threatened species (IUCN, 1994b) in comprehensive compilations by Collar *et al.* (1994). The categories of threat in decreasing order of severity are Critical (CR), Endangered (EN) and Vulnerable (VU). The criteria B, C, D1 and D2 are summarized in Table 8.3.

commonly by species qualifying by criteria B (limited range) and C (limited population size). Of all the species qualifying by criteria B and C, around 30% also qualified by A2. However, when the sample is restricted to species that qualify by A2 (the second histogram – decline total) the vast majority were in category C, and were therefore also listed because of their limited population size. The distribution of A2 and not-A2 species across criteria is significantly different from that expected from the overall distribution of species qualifying under different criteria ($\chi^2 = 42.2$, 3 df , $P < 0.0001$).

8.4.4 Discussion

For the bird species included in this analysis, limited population size was the most common reason for threatened listing; it was more important at higher threat levels and more common among species where future declines are expected to occur at a high rate. For birds at least, we can conclude that of the two kinds of rarity measured here, population size is more instrumental in listing than limited range size. In addition, if the threat categories provide a reasonably accurate assessment of impending extinctions, and if listing under the future decline rate criterion A2 is a

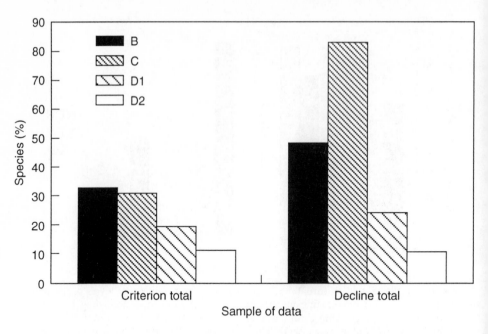

Figure 8.4 Percentage of threatened birds that are expected to decline by at least 20% over the next 10 years or three generations (whichever is the longer), that also meet the threshold values for small range sizes (B and D2) and small population size (C and D1). The criterion total shows the percentage of species that qualify for each of B (*n* = 685), C (*n* = 764), D1 (*n* = 351) or D2 (*n* = 256) that are also expected to be in future decline. The decline total shows the proportion of all declining species (*n* = 285) that also qualify for B, C, D1 and D2.

reliable indicator of future events, then it will be the species of limited population size that will be lost from the world's avifauna at the greatest rate.

A critical assumption here is that the non-inclusion of a criterion in the listing of threatened species suggests that the condition was not met, rather than that there were no data available to assess it. The data are of course incomplete. However, given the amount of work devoted to gathering reliable information on restricted-range birds at a global level (ICBP, 1992) and the generally high quality of available information (Bibby, 1995), population size data rather than range size data might be expected to be unavailable. Therefore the missing data would be more likely to bias the results towards the importance of limited range size rather than limited population size, and the bias would weaken rather than strengthen the results obtained here.

8.5 CONCLUSION

Throughout the analyses presented here, we have sought to differentiate the role of different aspects of rarity in the listing of threatened species as a measure of extinction risk. In the first place, we investigated aspects of rarity that most commonly lead to threatened listing at a relatively fine scale focusing on passerine birds within KwaZulu Natal, South Africa. Second, we undertook a global level analysis of all threatened birds, using specific information from quantitative criteria to investigate the characteristics of species estimated to be at highest extinction risk and most likely to suffer from severe declines in the short-term future. In both cases, the analyses suggested that rarity measured by small population size was most closely associated with extinction risk.

This result is unsurprising since population size and structure will govern population dynamics and will probably best reflect extinction risks (Soulé, 1987). Other variables, especially range size, are generally explicitly employed as surrogates because the information is more readily available, and because they are expected to provide a reasonable estimator of population size. Nevertheless, it is important to recognize that as surrogate measures become more removed from information directly affecting persistence, their power will decrease and will lead to errors that both overestimate and underestimate extinction risk. The poor relationship between habitat specificity and threatened status in the KwaZulu Natal data set is a case in point and illustrates that surrogate measures need to be evaluated at an appropriate scale and for the particular taxon and habitat in question.

There are some important caveats to these analyses. In the first place, we assume that threatened status provides a reliable estimator of extinction risk, and that the biases are similar between the different measures used to assess threatened status. Without doubt, the broad rules for assessing threatened status are very rough estimators of future extinction (Mace, 1994). However, there is no reason to suppose that they will be biased in a way that might influence the outcome of this analysis. In addition, it is reassuring that the two independent analyses here, using different methods at different scales, did produce similar results.

The conclusions we draw here may not be readily generalizable to other taxa. Relationships between rarity correlates differ among different major taxa (Gaston, 1994), and quite different associations might result from studies on, for example, invertebrates or lower plants. It is likely, though, that our results have some relevance for other large vertebrates that have broadly similar life histories to those of birds. Also, although we emphasize the importance of population size for extinctions driven by deterministic habitat loss, other more stochastic processes are likely to have been significant in the past and will be in the future. Catastrophic events causing rapid and extensive habitat change or directly affecting species within a local area are expected to play a major role in extinctions (Lande, 1993; Mangel and Tier, 1994). Here the major determinant of persistence is likely to be the occurrence of species outside the local area. We must assume that past extinctions caused by catastrophic events (introductions of alien predators or competitors, diseases, extreme climatic events, etc.) will have selectively eliminated species with localized distributions, rather than those with small local population sizes.

We conclude that rarity *per se* is not useful in the assessment or diagnosis of threatened status unless the precise way in which a species is rare can be determined. Particular population or distribution patterns contribute to increased extinction risk. Rarity might therefore be seen as a symptom or a correlate of a threatened species, but is an unreliable diagnostic feature. If the assumptions underlying the analyses here are reasonably robust, then we conclude that the set of rare species that are at highest risk of extinction, and which are declining most rapidly, are those with limited population size. All other things being equal, extinctions will bias the remaining set of rare species by reducing the number of species whose populations are always small, compared with reductions in other forms of rarity. Current anthropogenic activities result in habitat loss leading to fragmentation and reduction in total population numbers for an increasing array of species. Under these circumstances, we expect that the surviving set of species will be those that can persist in small areas because their behaviour, ecology or life history allows them to maintain large local abundance. Small body size, specialized habits favoured within their restricted range, good dispersal ability and the ability to occupy disturbed intermediate habitats are the kinds of factors that will be important, and that have probably also moulded past assemblages. With increasing habitat loss we expect to see a reduction in the numbers of rare species, especially those that are rare in terms of local abundance, and those that have only localized distributions. The set of 'rare' species may therefore become increasingly biased away from historical and recent patterns documented by Gaston (1994), and the balance between deterministic and catastrophic processes will be important in determining the form of this bias.

ACKNOWLEDGEMENTS

We would like to thank Steve Albon, Andrew Balmford and Peter Cotgreave for discussions and for commenting on a previous version of the manuscript. We are also grateful to Alison Stattersfield and Nigel Collar of BirdLife International for sharing data and ideas with us. GMM was supported by NERC during this work.

REFERENCES

Balmford, A. (1996) Extinction filters and current resilience: the significance of past selection pressures for conservation biology. *Trends in Ecology and Evolution*, **11**, 193–196.

Bibby, C.J. (1995). Recent past and future extinctions in birds, in *Extinction Rates* (eds J.H. Lawton and R.M. May), Oxford University Press, Oxford, pp. 98–110.

Bolger, D.T., Alberts, A.C. and Soulé, M.E. (1991) Occurrence patterns of bird species in habitat fragments: sampling, extinction and nested species subsets. *American Naturalist*, **137**, 155–166.

Brooke, R.K. (1984) *South African Red Data Book: Birds*, South African National Scientific Programmes Report, Pretoria.

Burbridge, A.A. and McKenzie, N.L. (1989) Patterns in the modern decline of Western Australia's native fauna: causes and conservation implications. *Biological Conservation*, **50**, 143–198.

Caughley, G. (1994) Directions in conservation biology. *Journal of Animal Ecology*, **63**, 215–244.

Collar, N.J., Crosby, M.J. and Stattersfield, A.J. (1994) *Birds to Watch 2 – The World List of Threatened Birds*, BirdLife International, Cambridge, UK.

Collar, N.J. and Stuart, S.N. (1985) *Threatened Birds of Africa and Related Islands. The IUCN/ICBP Red Data Book*, IUCN/ICBP, Cambridge, UK.

Cowling, R.M. and Bond, W.J. (1991) How small can reserves be? an empirical approach in Cape Fynbos, South Africa. *Biological Conservation*, **58**, 243–256.

Crawley, M.J. (1993) *GLIM for Ecologists*, Blackwell Scientific Publications, Oxford.

Cyrus, D.P., Robson, D. and Robson, N. (1980) *Bird Atlas of Natal*, University of Natal Press, Pietermaritzburg.

Diamond, J.M. (1984) 'Normal' extinctions of isolated populations, in *Extinctions* (ed. M.H. Nitecki), University of Chicago Press, Chicago, pp. 191–246.

Digby, P.G.N. and Kempton, R.A. (1987) *Multivariate Analysis of Ecological Communities*, Chapman & Hall, London.

Gaston, K.J. (1994) *Rarity*, Chapman & Hall, London.

Gilpin, M.E. and Soulé, M.E. (1986) Minimum viable populations: processes of species extinctions, in *Conservation Biology – the Science of Scarcity and Diversity* (ed. M.E. Soulé), Sinauer Associates, Michigan, pp. 19–34.

Given, D.R. and Norton, D.A. (1993) A multivariate approach to assessing threat and for priority setting. *Biological Conservation*, **64**, 57–66.

Goodman, D. (1987). The demography of chance extinction, in *Viable Populations for Conservation* (ed. M.E. Soulé), Cambridge University Press, Cambridge, pp. 11–34.

Happel, R.E., Noss, J.F. and Marsh, C.W. (1987) Distribution, abundance, and endangerment of primates, in *Primate conservation in the tropical rain forest* (eds C.W. Marsh, R.A. Mittermeier, and A.R. Liss) New York, pp. 63–82.

ICBP (1992) *Putting Biodiversity on the Map: priority areas for conservation*, International Council for Bird Preservation, Cambridge, UK.

IUCN (1994a) *1994 IUCN Red List of Threatened Animals*, IUCN, Gland, Switzerland.

IUCN (1994b) *IUCN Red List Categories*, IUCN, Gland, Switzerland.

Karr, J.R. (1982) Population variability and extinction in the avifauna of a tropical land bridge island. *Ecology*, **63**, 1975–1978.

Kennedy, M. (1992) *Australian Marsupials and Monotremes – an Action Plan for their Conservation*, IUCN, Gland, Switzerland.

Kunin, W.E. and Gaston, K.J. (1993) The biology of rarity: patterns, causes and consequences. *Trends in Ecology and Evolution*, **8**, 298–301.

Lande, R. (1988) Genetics and demography in biological conservation. *Science*, **241**, 1455–1460.

Lande, R. (1993) Risks of population extinction from demographic and environmental stochasticity and random catastrophes. *American Naturalist*, **142**, 911–927.

Laurance, W.F. (1991) Ecological correlates of extinction proneness in Australian tropical rain forest mammals. *Conservation Biology*, **5**, 80–89.

Lawton, J.H. (1995) Population dynamic principles, in *Extinction Rates* (eds J.H. Lawton and R.M. May). Oxford University Press, Oxford, UK.

Leigh, E.G. (1981) The average lifetime of a population in a varying environment. *Journal of Theoretical Biology*, **90**, 213–239.

Mace, G.M. (1994) Classifying threatened species: means and ends. *Philosophical Transactions of the Royal Society London B*, **344**, 91–97.

Mace, G.M. and Lande, R. (1991) Assessing extinction threats: toward a reevaluation of IUCN threatened species categories. *Conservation Biology*, **5**, 148–157.

Mangel, M. and Tier, C. (1994) Four facts every conservation biologist should know about persistence. *Ecology*, **75**, 607–614.

Mayr, E. (1963) *Animal Species and their Evolution*, Belknap Press, Cambridge, MA.

McCullagh, P. and Nelder, J.A. (1989) *Generalized Linear Models*, Chapman & Hall, London.

Munton, P. (1987) Concepts of threat to the survival of species used in Red Data Books and similar compilations, in *The Road to Extinction* (eds. R. Fitter and M. Fitter), IUCN, Gland, Switzerland, pp. 71–111.

Pimm, S.L., Jones, H.L. and Diamond, J.M. (1988) On the risk of extinction. *American Naturalist*, **132**, 757–785.

Rabinowitz, D. (1981) Seven forms of rarity, in *Biological Aspects of Rare Plant Conservation* (ed. H. Synge), Wiley, Chichester, pp. 205–217.

Rabinowitz, D., Cairns, S. and Dillon, T. (1986) Seven forms of rarity and their frequency in the flora of the British Isles, in *Conservation Biology: the Science of Scarcity and Diversity* (ed. M.E. Soulé), Sinauer Assoc. Inc, Sunderland, Mass., pp. 182–204.

Rabinowitz, D., Rapp, J.K., Cairns, S. and Mayer, M. (1989) The persistence of rare prairie grasses in Missouri: environmental variation buffered by reproductive output of rare species. *American Naturalist*, **134**, 525–544.

Richter-Dyn, N. and Goel, N.S. (1972) On the extinction of a colonising species. *Theoretical Population Biology*, **3**, 406–433.

Saetersdal, M. (1994) Rarity and species area relationships of vascular plants in deciduous woods, western Norway applications to nature reserve selection. *Ecography*, **17**, 23–38.

Simberloff, D. and Gotelli, N. (1984) Effects of insularization on plant species richness in the prairie forest ecotone. *Biological Conservation*, **29**, 27–46.

Sinclair, I., Hockey, P. and Tarbaton, N. (1993). *The Birds of Southern Africa*, Struick, Singapore.

Soulé, M.E. (1987) *Viable Populations for Conservation*, Cambridge University Press, Cambridge.

Terborgh, J. and Winter, B. (1980) Some causes of extinction, in *Conservation Biology: an evolutionary–ecological perspective* (eds M.E. Soulé and B.A. Wilcox), Sinauer Associates, Sunderland, Mass., pp. 119–133.

Thomas, C.D. (1991) Habitat use and geographic ranges of butterflies from the wet lowlands of Costa Rica. *Biological Conservation*, **55**, 269–281.

Thomas, C.D. and Mallorie, H.C. (1985) Rarity, species richness and conservation: butterflies of the Atlas Mountains in Morocco. *Biological Conservation*, **33**, 95–117.

WCMC (1992) *Global Biodiversity: Status of the Earth's Living Resources*, Chapman & Hall, London.

9 Population biology and rarity: on the complexity of density dependence in insect–plant interactions

William E. Kunin

9.1 INTRODUCTION

9.1.1 Density dependence and rarity

The subject of density dependence has been a central obsession of population ecologists for at least 60 years (e.g. Nicholson, 1933; Varley, 1947; Lack, 1954; Ricker, 1954; Orians, 1962; McLaren, 1971; Antonovics and Levin, 1980; Sinclair, 1989), and with good reason. If we could understand how the 'vital rates' of a population (the rates of reproduction and survivorship) change as it grows or shrinks, we could predict the population's dynamics. We should be able to calculate whether it will reach some stable equilibrium value, collapse to extinction, or even exhibit cyclical or chaotic dynamics. Yet past work on the subject has been remarkably one-sided. The study of density dependence concerns the consequences of a population being (at least locally) rare or common, yet the vast majority of the research to date has dealt with only half of this question. Almost all of the attention has been centred on the behaviour of populations at high density, as they approach or surpass the carrying capacity of their environments, and has focused on the processes that prevent species from becoming infinitely abundant. This is, of course, an important question, but it does not exhaust the subject of density-dependent effects. There are at least as many interesting issues to be studied on the opposite end of the spectrum, concerning the behaviour of populations as they approach local extinction. That is the focus of this chapter.

The Biology of Rarity. Edited by William E. Kunin and Kevin J. Gaston.
Published in 1997 by Chapman & Hall, London. ISBN 0 412 63380 9.

The theoretical population dynamics literature can serve as a case in point. A casual perusal of the literature (e.g. May *et al.*, 1974; May and Oster, 1976; Strong, 1986; Murray, 1993) produces a variety of models of density dependence in current usage (Figure 9.1). While they may differ slightly in form, all are in rough agreement about the consequences of high population densities. When it comes down to it, it is hard to argue against the notion that there is some maximum density beyond which over-population creates problems (although just how often species are close enough to the carrying capacity of their environments to experience such problems is an open question). Notice, however, that the models make very different assumptions about the consequences of life at low density. According to some models, the quality of life for the members of a species should improve indefinitely as the population grows increasingly rare, even at an accelerating pace; the rarer you are, the better your prospects. Others suggest that the vital rates of a low-density population may level off at some maximum value where density becomes irrelevant to success. Still other models invoke what is often termed 'inverse density-dependence' or

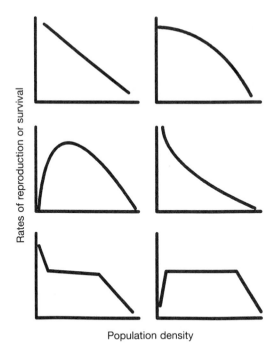

Population density

Figure 9.1 Density-dependent dynamics in various models culled from the theoretical literature. All agree that rates of survivorship and/or reproduction fall at high densities, although they differ somewhat in the precise pattern. The models differ greatly, however, in their assumptions about effects on moderate and low density populations. (Models are taken from May *et al.*, 1974; Strong, 1986; and Murray, 1993.)

'Allee effects' (after W.C. Allee who first noted the phenomenon; Allee *et al.*, 1949; Allee, 1951) at low density, in which the quality of life decreases drastically as the density of a population approaches zero. Thus, arguably, the major outstanding issue in modelling density dependence concerns the dynamics of rare populations.

Yet the behaviour of populations nearing extinction is of more than theoretical interest. One of the central issues in the rising discipline of conservation biology concerns the management of rare species populations. A better understanding of the dynamics of populations on the verge of extinction could have important practical applications for this effort. If the vital rates of rare populations improve as the populations shrink, there will be a negative feedback loop which could buffer the species' populations away from extinction. Conversely, if vital rates decline as the population grows smaller, a positive feedback loop is established which could cause a population to spiral towards extinction if it grows too rare. The nature of the density-dependent processes acting upon a species at low population density should thus tell us something about its vulnerability to extinction. If we could understand which populations (if any) were most likely to behave according to each of the models shown in Figure 9.1, we would be much better able to focus our conservation efforts productively on appropriate management techniques.

9.1.2 Types of rarity

Before proceeding further, however, a cautionary note is in order. So far, 'abundance' and 'rarity' have been written about as if there were simple, straightforward definitions for the terms. There are not. As a number of researchers have pointed out (e.g. Chapter 3; Rabinowitz, 1981; Fiedler and Ahouse, 1992; Gaston, 1994), rarity and abundance can be measured on a wide variety of different spatial scales, ranging from global range size down to local population density, resulting in a myriad of separate (but often interrelated) dimensions. Thus, when discussing the issue, it is important to be clear about precisely what sort of rarity one means to invoke. With very few exceptions, the density-dependent ecological interactions discussed here are influenced by abundance measured at a very local scale. The vital birth and death rates of a population are determined by the success or failure of its individuals in locating food and potential mates while avoiding predators, pathogens and competitors – interactions that are influenced by abundance only on the behavioural scales of the various species involved. Thus density, in the sense used by most students of density dependence, is near the bottom of the spectrum of abundance scales: the local abundance and spacing of individuals within a population. This is often correlated with abundance measured at other, larger, scales (e.g. Gaston, 1994) but the distinction is nonetheless important.

Even on a local scale, however, there are a number of nuances to be considered when considering rarity (Figure 9.2; see also Kareiva, 1983; Root and Kareiva, 1984). The most important components of local abundance are **population size** (the number of individuals in a population, **N**), **population density** (**D**, usually measured in terms of the inverse of spacing between neighbouring conspecifics) and **population purity** (**P**, the relative abundance of the species in the local mix – which depends in part on the density of the other species with which it is intermingled). These variables are easiest to define in populations of sessile organisms, where the precise spatial arrangement of individuals can be examined, but the distinctions may be important (if less amenable to study) in motile organisms as well. It should be pointed out that, like most abundance measures, these three axes are intrinsically interrelated. For example, where total population numbers are limited by space constraints, a high density population of a given species is almost automatically purer as well.

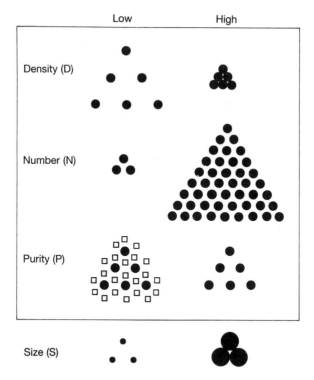

Figure 9.2 Various aspects of local population abundance. Density (D) reflects the spacing between individuals; number (N) reflects the size of the local population at that density; and purity (P) reflects the fraction of all local plants belonging to the focal species. Individual size (S) may both serve as an aspect of abundance, and be affected by other abundance factors.

Some of these factors also interact with a fourth variable, **individual size (S)**, a component of abundance not often considered but of potential importance in many species interactions. For example, in species with indeterminant growth, population density and purity can have strong effects on the size of individuals. Conversely, individual size (in species where it is roughly constant) helps to determine the maximum possible density of a population. Despite the confusing interactions between them, the effects of these four attributes of local abundance can often be teased apart by careful observation or experimental manipulation. The purpose of this chapter is to review how abundance on each of these scales (although it will focus largely on the first three) affects population interactions and performance.

Attention will be restricted here to insect–plant and plant–plant interactions and their consequences, primarily because they are easy subjects for experimental manipulation and study. As plants are sessile, their populations can be carefully measured or artfully arranged to allow the various components of local abundance to be scored or manipulated with a precision impossible in motile organisms. Insects, too, are ideal research subjects, as their behaviour and population dynamics often take place on a physical and temporal scale amenable to observation, and so their responses to particular arrangements of host plants can be observed easily. In particular, this chapter concentrates on two types of interactions: pollination and herbivory. The differences between them in their responses to different abundance patterns might be instructive. Many of the points made here could apply equally well to other systems – the purpose in the following review is to illustrate the richness of abundance-related changes in ecological interactions. It begins by exploring how each of the different aspects of local abundance affects pollination, then explores herbivory and takes a passing glance at competition as well. It then briefly examines the complexities produced when these processes are considered together. Finally, this is all related back to the central focus of this volume: the differences between rare and common species.

9.2 DENSITY DEPENDENCE IN VARIOUS PROCESSES

9.2.1 Pollination

Before examining the complexity of plant–pollinator interactions, perhaps we should begin by examining a conceptually simpler system. In wind pollination, gametes are dispersed passively from and to each plant, and consequently the cast of characters we need to understand is greatly reduced. Even so, density dependence in wind-pollinated plants seems likely to occur. The chance of a wind-borne pollen grain landing on a receptive stigma should be determined by the density of pollen grains per volume of air, which, in turn, should depend on the local density and (to a lesser extent) number of plants. Population purity should have little effect, except insofar as intervening plants interfere with air flow patterns, or filter the air in its passage.

There seem to be no studies examining the importance of population number or purity for wind pollination in the field, but there are a few on the effects of population density. Perhaps the best is by Allison (1990), who showed that pollination success in the Canadian yew (*Taxus canadensis*) is, indeed, inversely density dependent. Widely spaced individuals receive little pollen and set few seeds. Comparable results have been found in certain free-spawning marine animals (e.g. Levitan *et al*,. 1992; Levitan and Young, 1995).

The problem is more complicated with animal-pollinated plants. Here the success of a plant will depend on the quantity and quality of pollinator visits it can attract. Both of these factors are likely to depend on local abundance. Rathcke (1983) points out that the **quantity** of pollinator visits may vary in a complex way with local abundance. Where flowers are rare they may interact in a facilitative manner, by attracting pollinators to the area. At some point, however, increasingly common flowers may come to compete with one another for a limited pool of pollinators. Consequently the visitation rate per flower may rise and then fall as total floral density grows.

The **quality** of pollinator visits may have even more complex dynamics. It is determined by a number of factors, including the species mix of pollinators attracted, their morphology (e.g. hairiness), their behaviour when visiting a flower (and thus their probability of contacting anthers or stigmatic surfaces) and their movement patterns between flowers. This last factor is particularly sensitive to the density and spatial arrangement of floral populations, and thus may play a crucial role in determining the mating success of rare plants. To understand the quality of visits we must determine the degree to which pollinators will be 'flower constant' (moving between conspecific flowers and delivering pollen efficiently) or, altern-atively, behave as generalists and forage indiscriminately.

The value of a generalist pollinator depends greatly on the plant species' relative density. Generalist pollinators are very much like the wind in their pollen dispersal properties; they scatter pollen almost at random in the floral environment, and quickly cease to be effective pollinators at low relative density. (Note, however, that population purity, rather than density *per se*, is critical here; or more precisely the relative density of the species within the local mix of species whose flowers are simultaneously available to the pollinator in question.) Flower-constant pollinators, however, remain efficient pollen transfer agents even when conspecifics are widely scattered and interspersed with other species. By definition, such pollinators pass over all competing flower species and deliver pollen directly to appropriate targets. Unfortunately, from the plants' perspective, rare species (those with low numbers or density) make unprofitable targets for flower constancy, and so reductions in species abundance might result in shifts away from such behaviour.

But when **should** a pollinator be flower constant? Flower constancy may bring with it advantages in terms of improved flower handling, but it requires greater travel times, especially when the flower involved is rare. To analyse the decision even in a highly simplified context (with only two flower species, randomly interspersed, and identical in most respects) requires a fairly messy game-theoretic model, where the rewards from any particular strategy depend on the number of other foragers adopting it (Kunin and Iwasa, 1996). The model's behaviour suggests that pollinator constancy can be greatly influenced by floral densities (Figure 9.3). As the density of a plant species rises, the predicted mix of pollinators shifts from all generalists, to a mix of generalists and specialists, and finally to all specialists, with a resulting increase in pollination quality. Population purity among the flowers matters as well; if the competing plant in the

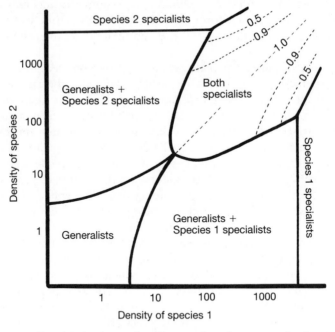

Figure 9.3 Predicted behaviour of pollinators foraging on a mixed array of two flower species. In this example, the two flower types are identical in reward characteristics. Pollinators are assumed to maximize net reward collection rates. These rates are affected by the behavioural mixture of foraging pollinators, so that the final mixture is determined in a game theoretic manner. Note that, as the density of any floral species falls, pollinators come to specialize disproportionately on their competitors. (Redrawn from Kunin and Iwasa, 1996.)

mixture grows more common, the generalists mentioned above are partially (or even completely) replaced by specialists on this other species, which are of even less value to the rare plant than are generalists. Overall, then, rare plants are caught between a qualitative rock and a quantitative hard place: they should have difficulty attracting specialist pollinators, but have little use for generalists. The message is clear: low absolute or relative density should result in pollination difficulties.

There have been a number of tests of density effects in pollination, but let us begin with an experiment on the wild mustard, *Brassica kaber* (Kunin, 1991, 1993). These self-incompatible annuals were planted out in fan-shaped density arrays at spacings ranging from 10.5 cm to 10 m apart, and the number and behaviour of pollinators that visited them were observed. In the outer rows of the fans, where density was lowest, the rate of pollinator visitation to the plants fell markedly. The quality of their visitors declined as well; where target plants were closely spaced, their insect visitors tended to move between conspecific flowers, but pollinators were much more likely to move on to other flower species where the focal species was sparsely planted in a mixed sward. The shift came about both because of a shift in the pollinator species visiting (from social bees and large syrphid flies to small syrphids and butterflies), and because of a behavioural shift within the pollinator species themselves. Honeybees, for example, which are renowned as flower-constant foragers, tended to behave as generalists where target flowers were rare. The result of low visitation rates and low pollinator quality was a marked decline in seed-set for the plants growing at low density.

When a very similar manipulation was performed on populations of the common ragwort, *Senecio jacobaea*, the effects on insect behaviour were almost identical (W.E. Kunin, in preparation) but in this case there was no effect on the plants' reproductive success. The reason, presumably, is that *S. jacobaea* is self-compatible, and therefore does not require efficient pollen transfer to set its seeds. Feinsinger *et al.* (1991) found a similar distinction when they manipulated floral densities of two species of cloud forest shrubs, one self-compatible and one self-incompatible. Pollinator behaviour was affected in both cases, but reproductive success changed only in the self-incompatible species. It should be noted, however, that these studies have looked only at the number of seeds set. If self-compatible species continue to set seed at low density, it may be by increasing self-pollination (Chapter 10). Thus the genetic makeup (and perhaps the fitness) of the seeds produced may well have been affected by density in the self-compatible species.

These patterns are repeated many times in studies on pollination and plant density (Table 9.1). The nearly universal result is that pollination is more difficult at low density – fewer or less effective pollinator visits are received, often leading to a decline in a flower's reproductive output. The exceptions are all self-compatible species where pollen transfer between

Table 9.1 Studies of plant population density (D) effects on pollination

Reference	Population type	Plant	Breeding system	Pollinator	Population density effects on	
					Pollination success	Reproductive success
Platt et al. (1974)	Natural (manipulated)	Astragalus canadensis	SI	Bees		+[1]
Silander (1978)	Natural	Cassia biflora		Bees	+?[1]	+[2]
Thomson (1981)	Natural	Potentilla gracilis		Flies and bees	+[2]	
		Senecio crassulus		Bees	+[2]	
Schmitt (1983)	Experimental	Senecio integerrimus	SC	Bees and butterflies	0[1]	0[1,3]
Feinsinger et al. (1986)	Natural	4 spp.	SC (all)	Hummingbirds	+ (2 spp.)[2] 0 (2 spp.)	
Klinkhamer et al. (1989)	Natural	Cynoglossum officinale	SI?	Bees	+[1]	
Klinkhamer and de Jong (1990)	Natural	Echium vulgare	SC	Bees	0[1]	
Allison (1990)	Natural	Taxus canadensis	SC	Wind	+[2]	+[1]
Feinsinger et al. (1991)	Experimental	Palicouria sp.	SI	Hummingbirds	+[2]	+[2]
		Besleria triflora	SC	Hummingbirds	+[2]	0[2]
Kunin (1992)	Natural	Diplotaxis erucoides	SI	Bees and flies		+[1,2,3]
House (1993)	Natural	Neolitsea dealbata	SI	Flies, beetles and bees	+[1,2]	+[1]
Kunin (1993)	Experimental	Brassica kaber	SI	Bees and flies	+[1]	+[1,2]
Kunin (in prep.)	Experimental	Senecio jacobaea	SC	Bees and flies	+[1]	0[1,3]
Total					+ 12 spp. 0 4 spp. − 0 spp.	+ 7 spp. 0 3 spp. − 0 spp.

SC = self-compatible; SI = self-incompatible.
Studies differed in methods used to measure pollination and reproductive success. For pollination success: 1 = pollinator visitation rate; 2 = number of pollen grains or pollen tubes. For reproductive success: 1 = fruit set; 2 = seeds per fruit; 3 = seeds per plant.
Note that pollination is nearly always found to be affected by plant density, and that reproductive success is affected as well as in all self-incompatible species.

individual plants is not required for setting seeds, and thus where pollinator quality is relatively unimportant. Self-compatibility, it seems, is an important modulating trait; it qualitatively shifts the nature of density dependence in pollination.

If we switch from population density to the study of population size, however, a rather different picture emerges. Triangular populations ranging in size from three to 78 wild mustard plants – a 26-fold range – were planted out; and pollinator visitation and seed-set on interior and edge individuals in each patch were monitored (W.E. Kunin, in preparation). Population density (the spacing between individuals) was held constant. The results were far from impressive: there was no significant difference in visitation rate as a function of patch size, although there was some slight difference as a function of placement within a patch. Not surprisingly, seed set was also unaffected by patch size. This result was confirmed by a follow-up study, this time using patches that differ not only in size, but also in density; population density had an important effect on pollination but population size again proved largely irrelevant.

Table 9.2 shows a number of other studies of population number effects on pollination. Almost all of these are observations of natural systems, so that population density is generally not controlled (and thus probably varies along with population size in many cases). There are some indications of rarity-related declines in reproductive success here: more species had positive numerical effects than negative ones. Fully half of the studies, however, (including all of the experimental studies) show no effect of population size on reproductive success. Population size effects, where they occur at all, seem to be much weaker than the density effects in pollination documented above.

To look at purity effects, let us return to the mustard density experiments described earlier. The density fans used in that experiment were planted in six different backgrounds: three of which had no competing flowers and three of which had a floral background of some other animal-pollinated plant (two of them dissimilar, and one quite similar to the focal plant in appearance and structure). Thus, for any given density on the array, we can compare results from pure and mixed floral stands. The effects of competing flowers on visitation rates were generally neutral or positive: dissimilar flowers had little or no effect on visitation to the low-density mustards, whereas similar-looking background flowers greatly increased visitation when the target plants were widely spaced. Competing flowers also had a big negative effect on pollinator quality. The presence of dissimilar flowers resulted in much lower visitor constancy at low density than was seen in plots without competing flowers. The effect was even more striking in plots where the competing flowers were similar to the target species; almost all visitors were inconstant. The result was a general decline in reproductive success at low density, but a decline which was very 'purity' dependent. In pure populations only slight declines were

Table 9.2 Studies of the effects of plant population size (N) on pollination

Reference	Patch type	Plant	Breeding system	Pollinator	Population size effects on		Problems
					Pollination success	Reproductive success	
Campbell (1985), Campbell and Motten (1985)	Experimental	*Stellaria pubera*	SC	Flies and bees	+[1]	0[4]	E
Sih and Baltus (1987)	Natural	*Nepeta cataria*	SC	Bees	+[1]	+[1]	A, B
Sowig (1989)	Natural	4 spp.	SC	Bees	− (2 spp.)[1] 0 (1 sp.)		A, C
Menges (1991)	Natural	*Silene regia*		Hummingbirds		+[2]	A, B, D
Lamont et al. (1993)	Natural	*Banksia goodii*	SI?	Birds and possums		+[3]	A, B, D
Van Treuren et al. (1993)	Natural Experimental	*Salvia pratensis*	SC	Bees	0[2] 0		B, C, E
Aizen and Feinsinger (1994)	Habitat fragments	16 spp.	SC: 6 spp. SI: 10 spp.	Various	+ (8 spp.)[3] 0 (7 spp.) − (1 sp.)	+ (6 spp.[1], 4 spp.[4]) 0 (7 spp., 9 spp.) − (2 spp., 1 sp.)	A, B, F
Kunin (in prep.)	Experimental	*Brassica kaber*	SI	Bees and flies	0[1]	0[1,4]	
Total					**+ 10 spp.** **0 10 spp.** **− 3 spp.**	**+ 8 spp.** **0 10 spp.** **− 1.5 spp.**	

SC = self-compatible; SI = self-incompatible.

For pollination: 1 = pollinator visitation rate; 2 = outcrossing rate; 3 = number of pollen grains or tubes. For reproductive success: 1 = fruit set; 2 = seed viability; 3 = seeds per plant; 4 = seeds per fruit.

Note that there is only weak evidence for reproductive effects, and that positive findings come exclusively from unmanipulated natural populations (where population size and density are often correlated).

Methodological problems which might have affected these results:

A = no experimental controls; B = reproductive failure may have been due to genetic differences, rather than population size *per se*; C = no measurement of reproductive performance; D = no pollination observations; E = little variation in population sizes; F = fragments may differ in environmental conditions; G = unnatural phenology of flowering.

Table 9.3 Studies of the effects of the purity (P) of plant populations on pollination

Reference	Population type	Plant	Breeding system	Pollinator	Competing plant	Population purity effects on		
						Pollination quantity	Pollination quality	Reproductive success
Waser (1978)	Experimental	Delphinium nelsoni		Hummingbirds	Ipomopsis aggregata			+[2]
Thomson (1981)	Natural	Potentilla gracilis		Flies and bees	P. fruticosa	−		
		Senecio crassulus		Bees	Helianthella + Helenium	−		
Campbell (1985), Campbell and Motten (1985)	Experimental (5 experiments)	Stellaria pubera	SC	Flies and bees	Claytonia virginica	−, 0, + (various exp'ts)	+[1,2]	+[1]
Feinsinger et al. (1991)	Experimental	Palicourea lasiorrachis	SI	Hummingbirds	Cephaelis elata	0	+[2]	+[2]
		Besleria triflora	SC	Hummingbirds	C. elata	0	0	+
Kunin (1993)	Experimental	Brassica kaber	SI	Bees and flies	B. hirta	−	+[1]	+[1,2]
					Eruca sativa	0	+	+
					Fagopyrum	0	+	+
Total						+ 0 spp. 0 5 spp. − 3 spp.	+ 5 spp. 0 1 spp. − 0 spp.	+ 7 spp. 0 0 spp. − 0 spp.

SC = self-compatible; SI = self-incompatible.
Pollination success is divided into pollination quantity (measured in all cases by pollinator visitation rate) and pollination quality. For pollination quality: 1 = pollinator visitation sequence (flower constancy); 2 = number of pollen grains or tubes. For reproductive success: 1 = fruit set; 2 = seeds per fruit. Note that, while population purity often decreased pollinator visitation rates, it almost always increased the quality of pollinator services and subsequent reproductive performance.

noted, and even then only at very low density. With competing flowers, however, density-related declines in performance were much stronger and came at much higher densities. This was especially true in the 'similar flowers' treatment, where seed set per fruit dropped to near zero when focal species plants were more than a metre or two apart.

Purity effects seem to be ubiquitous in the few studies that have been carried out to date (Table 9.3). Competing plants sometimes increase the quantity of visits, but they almost always decrease visit quality and reproductive success. Overall, pollination success grows with plant density, at least in self-incompatible plants; plant number has little effect, but plant population purity seems to have strong positive effects on reproduction.

9.2.2 Herbivory

Relative to pollination, a great deal of attention has already been paid in the literature to herbivory as it is affected by plant population patterns. Perhaps the best known statement on the subject comes from Root (1973) in his 'resource concentration hypothesis', which suggested disproportionate herbivore pressure on pure, large and dense plant populations. Similar ideas have been put forward in a number of other contexts. Janzen (1970) and Gillett (1962) suggested that seeds and seedlings growing near conspecific adult plants might suffer disproportionately high herbivory rates. Similarly, Tahvanainen and Root (1972) and Atsatt and O'Dowd (1976) stressed the possibility that heterospecific plants might confer a measure of 'associational resistance' to herbivores in mixed plant stands. The details of the models differ, but all agree on one point: rare plants (on any of our three local abundance scales) should be at an advantage when it comes to avoiding herbivory.

The data, however, have been much less unanimous. The ragwort density arrays discussed earlier provide a case in point. These plants are attacked by more than a dozen species of herbivorous insects, ranging in size from two species of tiny leaf-mining flies up through a series of flowerhead gallers and stem borers to the caterpillars of cinnabar moths (*Tyria jacobaeae*) which can defoliate whole plants. The effects of host density on these herbivores do not conform nicely to Root's theory. Among the leaf miners, one species clearly prefers low-density plantings while the other (which is less abundant) shows a statistically insignificant trend in the same direction. The same story holds for flowerhead gallers; the only species common enough to display a clear pattern appears strongly to prefer widely spaced plants to dense populations. On the other hand, none of the stem or root borers seem to be particularly influenced by density at all in choosing hosts. Of all the species studied, the only herbivore conforming to Root's predictions was the cinnabar moth, and even then only early instar caterpillars showed a high density bias. Late

Table 9.4 Studies of the effects of various aspects of local abundance on herbivory

Host plant population factor	Effects	Per unit area	Per unit plant	Per unit biomass
Density (D)[a]	+	**15**	5	5
	0	4	**9**	4
	−	4	**10**	7
Size (N)[b]	+	**16**		
	0	10		
	−	8		
Purity (P)[c]	+	**70**		
	0	14		
	−	24		

[a,b]Number of cases are updated from Kareiva (1983), with the following additional references: Density: Luginbill and McNeal, 1958; A'Brook, 1964; Way and Heathcote, 1966; Farrell, 1976; Finch and Skinner, 1976; Adesiyun, 1978; Mayse, 1978; Bach, 1980; Futuyma and Wasserman, 1980; McLain, 1981; MacKay and Singer, 1982; Evans, 1983; Root and Kareiva, 1984; Karban et al., 1989; Worthen, 1989; Segarra-Carmona and Barbosa, 1990; Brody, 1992; Kunin, unpublished data.
Number: Solomon, 1981; Maguire, 1983; Kareiva, 1985; Bach, 1986; Horton and Capinera, 1987; Bach, 1988a,b; Capman et al., 1990.
[c]Results from Andow (1991).

instar caterpillars (like all other herbivores showing any pattern at all) attacked low-density plants disproportionately.

The literature on herbivore responses to plant density is replete with such cases (e.g. Mayse, 1978; Bach, 1980; McLain, 1981; Karban et al., 1989): different herbivores can respond very differently to the same plant density array. This lack of clear pattern carries over into studies of population number as well. In natural populations of fireweed (*Chamerion angustifolium*), for example, two herbivore species preferentially attack large stands, one preferentially attacks small ones, and a fourth species doesn't seem to be affected either way (MacGarvin, 1982). Similarly in Bach's (1988a,b) experimental plantings of squash plants: one herbivore species attacks large stands disproportionately, one prefers small populations, and a third seems completely indifferent to host population size.

Overall, the patterns for herbivore effects in the published literature are much less clear-cut than those seen for pollination (Table 9.4). Looking first at density effects: the results depend critically on the statistic used to measure herbivore pressure. If this measure is herbivore density (that is, herbivores per unit area), there is a strong tendency for high-density plant populations to be attacked disproportionately. This is hardly surprising – there are more plants to eat per unit area in a dense population, and so it is understandable that they support a denser population of herbivores. Looking instead at herbivory per plant, we find a weak trend in the opposite direction, with greater attack rates on sparse stands – precisely the opposite of what Root predicted. However, this pattern may also be artefactual: the bias disappears when density-related differences in plant size are accounted for, as indicated by the lack of any clear pattern in studies of herbivory per unit of host biomass. The situation is not much clearer for herbivory as a function of host population size: large populations seem to face somewhat higher rates of herbivore attack, on average, than small ones, but the data are quite noisy, with quite a few counter-examples.

Of the three variables considered here, population purity seems to have the most consistent effect on herbivory. A fairly strong pattern seems to be emerging that pure species populations (monocultures) tend to suffer greater rates of herbivory than do mixed populations (reviewed by Andow, 1991). Even so, a significant number of cases show the opposite pattern, again suggesting that different herbivore species may respond differently to any given plant population pattern.

Despite the considerable attention that has been devoted to it, the pattern that emerges from herbivory studies is that the effects of host abundance are noisy and inconsistent. This is not to say that plant population patterns do not affect herbivore attack rates, but rather that they seem to affect different herbivores differently. There is, as of yet, no consensus answer as to what causes these differences; they may be due to different movement patterns, degrees of host specialization, sensory abilities, food requirements, competitive abilities, or vulnerability to predation. If we could understand what herbivore (and plant) traits modulate the different responses to plant population patterns, we would be much closer to understanding the nature of density dependence in herbivory.

9.2.3 A brief glance at competition

The bulk of this chapter has dealt with two types of density-related interactions: pollination and herbivory. If we are to understand the implications of abundance patterns for population performance, however, we would have to consider all aspects of a species' ecology, including many other species interactions. Competition is the most obvious missing ingredient, but an adequate description of abundance effects should consider a number of other processes as well (e.g. pathogens and mycorrhizal symbionts). There will be no attempt to cover competition adequately here, but the pattern of density effects in such interactions has been well documented (e.g. Antonovics and Levin, 1980; Pacala, 1989).

Intraspecific competition is intrinsically density dependent. At low densities, plants are unlikely to compete with conspecifics. As densities rise, however, first reproductive performance and then survival begin to drop markedly.

Similarly, interspecific competition is determined by the population density of other species in the mix or, in the terminology used here, by population purity. Holding conspecific density constant, the lower the purity of the target species' population, the more intensely its individuals will compete with heterospecifics, with negative (although not always significantly so) effects on vital rates. The effect of interspecific competition should vary greatly between species, depending on their relative competitive abilities. Thus for a competitively dominant species, the presence of other plants may have little effect, and so competition effects should depend on conspecific density only. For a competitive subordinate, however, the presence of heterospecifics might be as harmful, or even more harmful, than the presence of conspecifics, leading to very strong population purity effects. The effects of population number on competition, on the other hand, should be fairly negligible in all cases, except at very small population sizes where the lack of competition at the edges of a population may be important.

9.3 PUTTING THE PIECES TOGETHER

Combining all these effects can make for some interesting patterns. We can view the three components of local abundance discussed above (density, purity and number) as the three dimensions of a cube. Figure 9.4 illustrates the general patterns of pollination, herbivory and competition effects on fitness (as reviewed above) in such cubes, with the shading of the balls filling each cube reflecting the severity of the effect. Thus pollination (Figure 9.4a), at least in the self-incompatible case illustrated, is strongly affected by population density and purity, but only weakly influenced by population size, resulting in dark shading along the lower back left edge of the cube (where both density and purity are lowest). Herbivory, on the other hand, is more influenced by the size and purity of host population than by density (Figure 9.4b), reflected by the dark shading of the front right-hand vertical edge (recall, however, that effects seem to vary greatly among cases, and so different species would produce markedly different graphs). Finally, competition (Figure 9.4c) is almost entirely dependent on density (for intraspecific competition) and purity (for interspecific effects), and so is sharply focused in the upper back left edge of the cube.

Combining these components of performance results in a rather complex pattern of fitness (Figure 9.4d). Much of the volume of the cube is now darkened, leaving a lighter band of relatively high fitness stretching from the lower front corner (with low numbers and density but high purity) into the interior of the cube (moderate levels of all three variables). The picture is dramatically different if we assume the plant to be reproductively self-compatible, and thus largely immune to abundance-related pollination problems. In this case (Figure 9.4e), population performance is considerably higher, especially in the lower back edge (where both density and purity are low). In a similar way, plant species with different patterns of

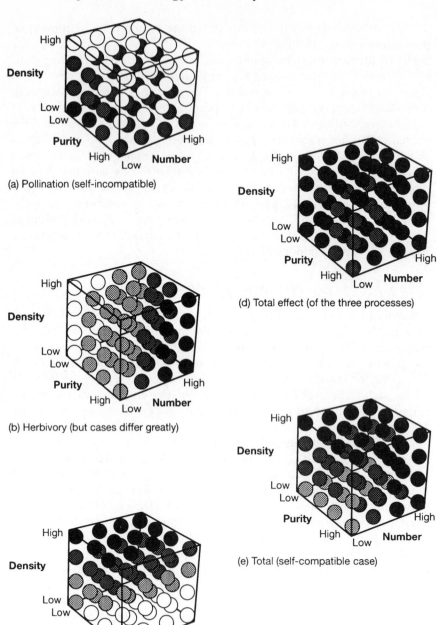

(a) Pollination (self-incompatible)

(b) Herbivory (but cases differ greatly)

(c) Competition (inter + intra-specific)

(d) Total effect (of the three processes)

(e) Total (self-compatible case)

Figure 9.4 Effects of the density, number and purity of individuals in plant populations on fitness, as influenced by the various interactions discussed in this chapter, taken separately and combined. Darker shading indicates decreased

competition effects (due to higher or lower competitive dominance) or of herbivory effects (due to differences in whatever traits modulate herbivore attack patterns) could display markedly different overall fitness patterns from those shown. If we were to add other ecological interactions – pathogens and mycorrhizae, for example – the results could be even more complex.

There are several important points that these figures are meant to underscore.

1. **Abundance is critically important** in a number of significant ecological interactions. All three of the processes explored here are greatly influenced by at least some aspects of local population patterns.
2. **Details of distribution matter.** Population density, number and purity may have very different implications for any given process, and different processes seem to respond differently to any given abundance variable.
3. **Rare and common species are likely to face very different ecological pressures.** Even where their fitness is similar, species with very different positions in 'abundance space' are likely to face markedly different sets of problems. Thus low-density populations face pollination difficulties, whereas high-density ones face stiff intraspecific competition; low purity populations face a mixture of pollination and interspecific competition problems, while high purity populations are often subject to significant herbivory pressure.
4. **Some simple diagnostic traits may modulate abundance-related effects.** Thus, for example, a plant's breeding system determines its vulnerability to density-related pollination problems, and consequently the overall form of its abundance response. Similarly, the competitive dominance of a species will influence the shape of its competition effect, with dominant species having little response to population purity and subordinate species having strong purity effects. An important outstanding issue is: what traits modulate herbivore abundance responses, to explain the diametrically opposite patterns displayed by different herbivores of the same plant populations?

---

fitness. (a) Pollination difficulties occur primarily where population density and purity are low – the back lower left edge of the cube. (b) Herbivory patterns differ greatly between species, but tend to show the highest loads in populations with high purity and large numbers of individuals – the front left edge. (c) Competition is most intense where population density is high and population purity is low – the upper back left edge of the figure. (d) Combining these effects results in a nearly uniform reduction of fitness, with different population patterns facing different ecological pressures. (e) In self-compatible plants, however, the fitness implications of pollination are removed, leading to fitness advantages in sparse populations of low purity. In similar ways, the fitness landscape would vary with differing herbivory patterns or levels of competitive dominance.

9.4 POPULATION BIOLOGY AND RARE–COMMON DIFFERENCES

All of this is well and good, but how does it relate to the main subject of this book – the explanation of rare–common differences? There are two primary reasons for including a consideration of population biology and species interactions here. First of all, there is a growing list of examples of what might be considered fundamental species traits which are, in fact, plastic. In many of these cases, the ecological processes that drive these trait changes are themselves density related. For example, an increase in local population density or number would result in an increasingly wide range of habitat utilization in a species following Fretwell's (1972; Fretwell and Lucas, 1969) 'Ideal Free Distribution' (see also Rosenzweig, 1981). Similarly, so-called 'optimal diet' or 'prey' models (e.g. Schoener, 1971; Charnov and Orians, 1973) suggest that a species' diet should shift from narrow specialization to increasingly broad generalization as resources become scarce, a prediction that has been supported by dozens of empirical studies (reviewed by Stephens and Krebs, 1986). Large shifts in the abundance of a species could easily result in the sorts of changes in prey encounter rates that drive these models, resulting in abundance-dependent shifts in a species' diet breadth. Conversely, a population change in a prey species might change the prey encounter rate experienced by its predators, resulting in an abundance-related shift in predation pressure. Where species display inducible defences, these chemical or physical properties could easily come to shift with abundance-related changes in attack rates and resources (Karban *et al.*, 1989). Density- or purity-related shifts in pollinator visitation can result in differing pollinator assemblages (Schmitt, 1983; Kunin, 1993), shifts in floral phenology (Stanton *et al.*, 1987; Lyons and Mully, 1992), shifts in reproductive allocation strategies (Paige and Whitham, 1987), or changes in floral sex ratios (Smith, 1981). Shifts in local population density can even affect the social systems of animals; the Seychelles warbler (*Acrocephalus sechellensis*), for example, shifts from defending individual territories to group breeding as population densities rise (Kombeur, 1992; Mumme, 1992). Such density-dependent ecological changes could be a contributing factor to the sets of rare–common differences which are the focus of this volume.

There is, however, a second and more general reason for concerning ourselves with the abundance-related ecological shifts that have been the subject of most of this chapter. The peculiar set of pressures outlined above is the basis for many of the transformations that are discussed elsewhere in this volume. They are (to paraphrase Hutchinson, 1965) the ecological stage upon which the saga of selective extinction (Chapters 7 and 8), genetic shifts (Chapter 10) and evolutionary change (Chapters 11 and 12) are played. If there is selective pressure for adaptation by rare species, it is largely because rarity affects the vital rates of the populations involved. If species with a biased set of traits are likely to become extinct in the face of rarity, it may well be due to the ways in which those traits (e.g. self-compatibility) modulate the implications of abundance.

REFERENCES

A'Brook, J. (1964) The effect of planting date and spacing on the incidence of groundnut rosette disease and of the vector, *Aphis craccivora* Koch, at Mokwa, northern Nigeria. *Annals of Applied Biology*, **54**, 199–208.

Adesiyun, A.A. (1978) Effects of seeding density and spatial distribution of oat plants on colonization and development of *Oscinella frit* (Diptera: Chloropidae). *Journal of Applied Ecology*, **15**, 797–808.

Aizen, M.A. and Feinsinger, P. (1994) Forest fragmentation, pollination, and plant reproduction in a Chaco dry forest, Argentina. *Ecology*, **75**, 330–351.

Allee, W.C. (1951) *The Social Life of Animals*, 2nd edn, Beacon Press Boston (1958 reprint).

Allee, W.C., Emerson, A.E., Park, O. (1949) *Principles of Animal Ecology*, W.B. Saunders Company, Philadelphia.

Allison, T.D. (1990) Pollen production and plant density affect pollination and seed production in *Taxus canadensis*. *Ecology*, **71**, 516–522.

Andow, D.A. (1991) Vegetational diversity and arthropod population response. *Annual Review of Entomology*, **36**, 561–586.

Antonovics, J. and Levin, D.A. (1980) The ecological and genetic consequences of density-dependent regulation in plants. *Annual Review of Ecology and Systematics*, **11**, 411–452.

Atsatt, P.R. and O'Dowd, D.J. (1976) Plant defense guilds. *Science*, **193**, 24–29.

Bach, C.E. (1980) Effects of plant density and diversity on the population dynamics of a specialist herbivore, the striped cucumber beetle, *Acalymma vittata* (Fab.). *Ecology*, **61**, 1515–1530.

Bach, C.E. (1986) A comparison of the responses of two tropical specialist herbivores to host plant patch size. *Oecologia* (Berl.), **68**, 580–584.

Bach, C.E. (1988a) Effects of host plant patch size on herbivore density: patterns. *Ecology*, **69**, 1090–1102.

Bach, C.E. (1988b) Effects of host plant patch size on herbivore density: underlying mechanisms. *Ecology*, **69**, 1103–1117.

Brody, A.K. (1992) Oviposition choices by a pre-dispersal seed predator (*Hylemya* sp.). I. Correspondence with hummingbird pollinators, and the role of plant size, density and floral morphology. *Oecologia* (Berl.), **91**, 56–62.

Campbell, D.R. (1985) Pollinator sharing and seed set of *Stellaria pubera*: competition for pollination. *Ecology*, **66**, 544–553.

Campbell, D.R. and Motten, A.F. (1985) The mechanism of competition for pollination between two forest herbs. *Ecology*, **66**, 554–563.

Capman, W.C., Batzli, G.O. and Simms, L.E. (1990) Responses of the common sooty wing skipper to patches of host plants. *Ecology*, **71**, 1430–1440.

Charnov, E.L. and Orians, G.H. (1973) *Optimal Foraging: some theoretical explorations*, self-published MS, Seattle, WA.

Evans, E.W. (1983) The influence of neighboring hosts on colonization of prairie milkweeds by a seed-feeding bug. *Ecology*, **64**, 648–653.

Farrell, J.A.K. (1976) Effects of groundnut crop density on the population dynamics of *Aphis craccivora* Koch (Hemiptera, Aphididae) in Malawi. *Bulletin of Entomology Research*, **66**, 317–329.

Feinsinger, P., Murray, K.G., Kinsman, S. and Busby, W.H. (1986) Floral neighborhood and pollination success in four hummingbird-pollinated cloud forest plant species. *Ecology*, **67**, 449–464.

Feinsinger, P., Tiebout, H.M., III and Young, B.E. (1991) Do tropical bird-pollinated plants exhibit density-dependent interactions? Field experiments. *Ecology*, **72**, 1953–1963.

Fiedler, P.L. and Ahouse, J.J. (1992) Hierarchies of cause: towards an understanding of rarity in vascular plant species, in *Conservation Biology: the theory and practice of nature conservation, preservation, and management* (eds P.L. Fiedler and S.K. Jain), Chapman & Hall, New York, pp. 23–47.

Finch, S. and Skinner, G. (1976) The effect of plant density on populations of the cabbage root fly (*Erioischia brassicae* (Bch.)) and the cabbage stem weevil (*Ceutorhynchus quadridens* (Panz.)) on cauliflowers. *Bulletin of Entomology Research*, **66**, 113–123.

Fretwell, S.D. (1972) *Populations in a Seasonal Environment*, Princeton University Press, Princeton, New Jersey.

Fretwell, S.D. and Lucas, H.L., Jr (1969) On territorial behaviour and other factors influencing habitat distribution in birds. I. Theoretical developments. *Acta Biotheoretica*, **19**, 16–36.

Futuyma, D.J. and Wasserman, S.S. (1980) Resource concentration and herbivory in oak forests. *Science*, **210**, 920–922.

Gaston, K.J. (1994) *Rarity*, Chapman & Hall, London.

Gillett, J.B. (1962) Pest pressure, an underestimated factor in evolution. *Systematics Association, publication number 4, Taxonomy and Geography* (April), 37–46.

Horton, D.R. and Capinera, J.L. (1987) Effects of plant diversity, host density, and host size on population ecology of the Colorado potato beetle (Coleoptera: Chrysomelidae). *Environmental Entomology*, **16**, 1019–1026.

House, S.M. (1993) Pollination success in a population of dioecious rain forest trees. *Oecologia* (Berl.) **96**, 555–561.

Hutchinson, G.E. (1965) *The Ecological Theater and the Evolutionary Play*, Yale University Press, New Haven.

Janzen, D.H. (1970) Herbivores and the number of tree species in tropical forests. *The American Naturalist*, **104**, 501–528.

Karban, R., Brody, A.K. and Schnathorst, W.C. (1989) Crowding and a plant's ability to defend itself against herbivores and diseases. *The American Naturalist*, **134**, 749–760.

Kareiva, P. (1983) Influence of vegetation texture on herbivore populations: resource concentration and herbivore movement, in *Variable Plants and Herbivores in Natural and Managed Systems* (eds R.F. Denno and M.S. McClure), Academic Press, Inc., New York, pp. 259–289.

Kareiva, P. (1985) Finding and losing host plants by *Phyllotreta*: patch size and surrounding habitat. *Ecology*, **66**, 1809–1816.

Klinkhammer, P.G.L. and de Jong, T.J. (1990) Effects of plant size, plant density and sex differential nectar reward on pollinator visitation in the protandrous *Echium vulgare* (Boraginaceae). *Oikos*, **57**, 399–405.

Klinkhamer, P.G.L., de Jong, T.J. and de Bruyn, G.-J. (1989) Plant size and pollinator visitation in *Cynoglossum officinale*. *Oikos*, **54**, 201–204.

Kombeur, J. (1992) Importance of habitat saturation and territory quality for the evolution of cooperative breeding in the Seychelles warbler. *Nature*, **358**, 493–495.

Kunin, W.E. (1991) Few and far between: plant population density and its effects on insect–plant interactions. PhD thesis, University of Washington.

Kunin, W.E. (1992) Density and reproductive success in wild populations of *Diplotaxis erucoides* (Brassicaceae). *Oecologia* (Berl.), **91**, 129–133.

Kunin, W.E. (1993) Sex and the single mustard: population density and pollinator behavior effects on seed-set. *Ecology*, **74**, 2145–2160.

Kunin, W.E. and Iwasa, Y. (1996) Pollinator foraging strategies in mixed floral arrays: density effects and floral constancy. *Theoretical Population Biology*, **49**, 232–263.

Lack, D. (1954) *The Natural Regulation of Animal Numbers*, Oxford University Press, New York.

Lamont, B.B., Klinkhamer, P.G.L. and Witkowski, E.T.F. (1993) Population fragmentation may reduce fertility to zero in *Banksia goodii* – a demonstration of the Allee effect. *Oecologia* (Berl.), **94**, 446–450.

Levitan, D.R. and Young, C.M. (1995) Reproductive success in large populations – empirical measures and theoretical predictions of fertilization in the sea biscuit *Clypeaster rosaceus*. *Journal of Experimental Marine Biology and Ecology*, **190**, 221–241.

Levitan, D.R., Sewall, M.A. and Chia, F.S. (1992) How distribution and abundance influence fertilization success in the sea urchin *Strongylocentrotus franciscanus*. *Ecology*, **73**, 248–254.

Luginbill, P., Jr and McNeal, F.H. (1958) Influence of seeding density and row spacings on the resistance of spring wheats to the wheat stem sawfly. *Journal of Economic Entomology*, **51**, 804–808.

Lyons, E.E. and Mully, T.W. (1992) Density effects on flowering phenology and mating potential in *Nicotiana alata*. *Oecologia (Berl.)*, **91**, 93–100.

MacGarvin, M. (1982) Species–area relationships of insects on host plants: herbivores on rosebay willowherb. *Journal of Animal Ecology*, **51**, 207–223.

MacKay, D.A. and Singer, M.C. (1982) The basis of an apparent preference for isolated host plants by ovipositing *Euptychia libye* butterflies. *Ecological Entomology*, **7**, 299–303.

Maguire, L.A. (1983) Influence of collard patch size on population densities of lepidopteran pests (Lepidoptera: Pieridae, Plutellidae). *Environmental Entomology*, **12**, 1415–1419.

May, R.M., Conway, G.R., Hassell, M.P. and Southwood, T.R.E. (1974) Time delays, density-dependence, and single species oscillations. *Journal of Animal Ecology*, **43**, 747–770.

May, R.M. and Oster, G.F. (1976) Bifurcations and dynamic complexity in simple ecological models. *The American Naturalist*, **110**, 573–599.

Mayse, M.A. (1978) Effects of spacing between rows on soybean arthropod populations. *Journal of Applied Ecology*, **15**, 439–450.

McLain, D.K. (1981) Resource partitioning by three species of hemipteran herbivores on the basis of host plant density. *Oecologia* (Berl.), **48**, 414–417.

McLaren, I.A. (1971) *Natural Regulation of Animal Numbers*, Atherton, New York.

Menges, E.S. (1991) Seed germination percentage increases with population size in a fragmented prairie species. *Conservation Biology*, **5**, 158–164.

Mumme, R.L. (1992) Delayed dispersal and cooperative breeding in the Seychelles warbler. *Trends in Ecology and Evolution*, **7**, 330–331.

Murray, J.D. (1993) *Mathematical Biology* (Biomathematics series, 19), 2nd edn, Springer, Berlin.

Nicholson, A.J. (1933) The balance of animal populations. *Journal of Animal Ecology*, **2**, 132–178.

Orians, G.H. (1962) Natural selection and ecological theory. *The American Naturalist*, **96**, 257–263.

Pacala, S.W. (1989) Plant population dynamic theory, in *Perspectives in Ecological Theory* (eds J. Roughgarden, R.M. May and S.A. Levin), Princeton University Press, Princeton, pp. 54–67.

Paige, K.N. and Whitham, T.G. (1987) Flexible life history traits: shifts by scarlet gilia in response to pollinator abundance. *Ecology*, **68**, 1691–1695.

Platt, W.J., Hill, G.R. and Clark, S. (1974) Seed production in a prairie legume (*Astragalus canadensis* L.): interactions between pollination, predispersal seed predation, and plant density. *Oecologia* (Berl.), **17**, 55–63.

Rabinowitz, D. (1981) Seven forms of rarity, in *The Biological Aspects of Rare Plant Conservation* (eds H. Synge), John Wiley and Sons Ltd, pp. 205–217.

Rathcke, B. (1983) Competition and facilitation among plants for pollination, in *Pollination Biology* (ed. L. Real), Academic Press, Inc., Orlando, FL, pp. 305–329.

Ricker, W.E. (1954) Stock and recruitment. *Journal of the Fisheries Research Board of Canada*, **11**, 559–623.

Root, R.B. (1973) Organization of a plant–arthropod association in simple and diverse habitats: the fauna of collards (*Brassica oleracea*). *Ecological Monographs*, **43**, 95–124.

Root, R.B. and Kareiva, P.M. (1984) The search for resources by cabbage butterflies (*Pieris rapae*): ecological consequences and adaptive significance of Markovian movements in a patchy environment. *Ecology*, **65**, 147–165.

Rosenzweig, M.L. (1981) A theory of habitat selection. *Ecology*, **62**, 327–335.

Schmitt, J. (1983) Flowering plant density and pollinator visitation in *Senecio*. *Oecologia* (Berl.), **60**, 97–102.

Schoener, T.W. (1971) Theory of feeding strategies. *Annual Review of Ecology and Systematics*, **2**, 369–404.

Segarra-Carmona, A. and Barbosa, P. (1990) Influence of patch plant density on herbivory levels by *Etiella zinkenella* (Lepidoptera: Pyralidae) on *Glycine max* and *Crotalaria pallida*. *Environmental Entomology*, **19**, 640–647.

Sih, A. and Baltus, M. (1987) Patch size, pollinator behavior, and pollinator limitation in catnip. *Ecology*, **68**, 1679–1690.

Silander, J.A., Jr (1978) Density-dependent control of reproductive success in *Cassia biflora*. *Biotropica*, **10**, 292–296.

Sinclair, A.R.E. (1989) Population regulation in animals, in *Ecological Concepts: the contribution of ecology to an understanding of the natural world* (ed. J.M. Cherrett), Blackwell Scientific Publications, Oxford, pp. 179–241.

Smith, C.C. (1981) The facultative adjustment of sex ratio in Lodgepole pine. *The American Naturalist*, **118**, 297–305.

Solomon, B.P. (1981) Response of a host-specific herbivore to resource density, relative abundance, and phenology. *Ecology*, **62**, 1205–1214.

Sowig, P. (1989) Effects of flowering plant's patch size on species composition of pollinator communities, foraging strategies, and resource partitioning in bumblebees (Hymenoptera: Apidae). *Oecologia* (Berl.), **78**, 550–558.

Stanton, M.L., Bereczky, J.K. and Hasbrouck, H.D. (1987) Pollination thoroughness and maternal yield regulation in wild radish, *Raphanus raphanistrum* (Brassicaceae). *Oecologia* (Berl.), **74**, 68–76.

Stephens, D.W. and Krebs, J.R. (1986) *Foraging Theory*, Princeton University Press, Princeton, NJ (Monographs in Behavior and Ecology).

Strong, D.R. (1986) Density-vague ecology and liberal population regulation in insects, in *Community Ecology* (eds J. Diamond and T.J. Case), Harper and Row, New York, pp. 313–327.

Tahvanainen, J.O. and Root, R.B. (1972) The influence of vegetational diversity on the population ecology of a specialized herbivore, *Phyllotreta cruciferae* (Coleoptera: Chrysomelidae). *Oecologia* (Berl.), **10**, 321–346.

Thomson, J.D. (1981) Spatial and temporal components of resource assessment by flower-feeding insects. *Journal of Animal Ecology*, **50**, 49–59.

Van Treuren, R., Bijlsma, R., Ouborg, N.J. and Van Delden, W. (1993) The effects of population size and plant density on outcrossing rates in locally endangered *Salvia pratensis*. *Evolution*, **47**, 1094–1104.

Varley, G.C. (1947) The natural control of population balance in the knapweed gall-fly (*Urophora jaceana*). *Journal of Animal Ecology*, **16**, 139–187.

Waser, N.M. (1978) Competition for hummingbird pollination and sequential flowering in two Colorado wildflowers. *Ecology*, **59**, 934–944.

Way, M.J. and Heathcote, G.D. (1966) Interactions of crop density of field beans, abundance of *Aphis fabae* Scop., virus incidence and aphid control by chemicals. *Annals of Applied Biology*, **57**, 409–423.

Worthen, W.B. (1989) Effects of resource density on mycophagous fly dispersal and community structure. *Oikos*, **54**, 145–153.

10 Genetic consequences of different patterns of distribution and abundance

Jeffrey D. Karron

10.1 INTRODUCTION

The term 'rarity' has been applied to species with diverse patterns of distribution and abundance (Chapter 3; Drury, 1974, 1980; Harper, 1981; Rabinowitz, 1981; Fiedler and Ahouse, 1992; Kunin and Gaston, 1993; Gaston, 1994). For example, localized endemics have small geographical distributions but often occur at high population density, whereas sparse species occur at low density but frequently have broad ranges (Rabinowitz *et al.*, 1986). Since evolutionary processes are influenced by the size and density of populations (Wright, 1969; Barrett and Kohn, 1991; Ellstrand and Elam, 1993), we may expect to find differences in the genetic structure of distinct forms of rare species.

This chapter explores the evolutionary consequences of different types of rarity. The review begins with a brief summary of relevant evolutionary theory. It then presents results from recent studies that have quantified patterns of genetic diversity and breeding systems in rare vascular plants. The chapter concludes with a description of ongoing experimental studies concerning the effects of population density on mating patterns in monkeyflower, *Mimulus ringens*.

10.2 EVOLUTIONARY CONSEQUENCES OF SMALL POPULATION SIZE

The evolutionary potential of a species may be influenced by the amount and distribution of genetic variation within and among populations

The Biology of Rarity. Edited by William E. Kunin and Kevin J. Gaston.
Published in 1997 by Chapman & Hall, London. ISBN 0 412 63380 9.

(Frankel and Soulé, 1981). Genetic diversity facilitates establishment in new habitats and persistence under changing environmental conditions (Beardmore, 1983; Huenneke, 1991). Several studies have demonstrated a genetic basis for phenotypic differences among populations growing in distinct habitats (Clausen *et al.*, 1940, 1948; McNeilly and Bradshaw, 1968; Van Dijk, 1989). In addition, reciprocal transplant experiments have shown that resident genotypes have a significant fitness advantage over non-resident genotypes (Chapin and Chapin, 1981; McGraw and Antonovics, 1983; Schmidt and Levin, 1985). Within populations, genetic variation allows adaptive responses to fine-scale environmental hetero-geneity (Linhart, 1974; Silander, 1979) and resistance to disease epidemics (Burdon, 1987).

Natural catastrophes, habitat fragmentation and changing environ-mental conditions may dramatically reduce the size of rare plant and animal populations (Menges, 1991). When a population undergoes a bottleneck (a sharp reduction in number of individuals), rare alleles may be lost, resulting in a decrease in allelic diversity. Newly-formed populations colonized by a small number of individuals are also likely to have low diversity because the founders may carry only a subset of the alleles of the parental population (Barrett and Kohn, 1991). In both cases, the reduction in heterozygosity is often minimal if the population rebounds rapidly in size (Nei *et al.*, 1975; Barrett and Kohn, 1991).

Populations that remain small (e.g. < 50 breeding individuals) for several generations are particularly susceptible to **genetic drift** (random changes in the frequency of neutral alleles from one generation to the next). These sampling effects, which are cumulative over time, may lead to reduced allelic diversity and heterozygosity within populations. Genetic drift may also increase the amount of genetic differentiation among populations (Wright, 1969; Barrett and Kohn, 1991; Ellstrand and Elam, 1993).

Small populations frequently have increased homozygosity due to the effects of biparental inbreeding. In addition, self-compatible hermaphro-ditic individuals may self-fertilize – the most extreme form of inbreeding. The frequency of selfing tends to increase when outcross mates are scarce, as is the case in very small or low-density populations (Baker, 1955; Jain, 1976; Jarne and Charlesworth, 1993).

Inbreeding may have important consequences for individual and population fitness. Randomly mating populations often harbour a large genetic load (a gene pool of lethal or highly deleterious recessive alleles). When populations that have historically mated at random begin inbreeding, these deleterious alleles are exposed to selection in homozygous indi-viduals (Wright, 1977). Over time, inbreeding and selection should purge a population of much of its genetic load, resulting in lower levels of inbreeding depression (Lacy, 1992).

10.3 PATTERNS OF GENETIC DIVERSITY IN RARE PLANT SPECIES

Since genetic drift and population bottlenecks reduce allelic diversity in small populations, levels of selectively neutral (or nearly neutral) genetic variation should be positively correlated with population size. A positive relationship between allozyme polymorphism and population size has been documented in several species (Ellstrand and Elam, 1993), including *Salvia pratensis* (Figure 10.1; Van Treuren *et al.*, 1991). Very few studies have investigated the relationship between population size and polymorphism for phenotypic traits that are influenced by selection. Husband and Barrett (1992) determined the number of style morphs in 167 populations of tristylous *Eichhornia paniculata*. Populations with three style morphs were significantly larger than those containing one or two morphs, suggesting that genetic drift may lead to the loss of style morphs in small populations.

Localized endemic species, which frequently consist of a few relatively small populations, are highly susceptible to drift. This prediction was tested by comparing levels of allozyme polymorphism in ten sets of geographically restricted species and widespread congeners (Karron, 1987a). Paired comparisons increase the probability that the restricted and widespread species have similar ecological features and phylogenetic histories (Karron *et al.*, 1988). This is desirable since levels of polymorphism are known to be correlated with life history characteristics

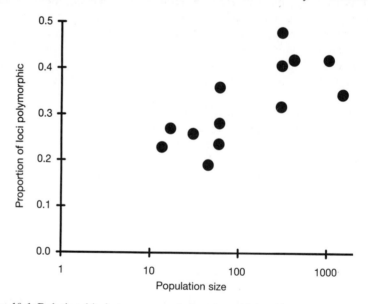

Figure 10.1 Relationship between population size and genetic polymorphism in 13 populations of *Salvia pratensis* ($r = 0.619$). (Modified from Van Treuren *et al.*, 1991.)

(Hamrick *et al.*, 1979; Hamrick, 1983; Loveless and Hamrick, 1984). As shown in Figure 10.2, genetic variation in the localized endemics is significantly lower than levels of polymorphism in the widespread congeners.

A similar result was obtained by Hamrick and Godt (1990), who compiled data on levels of polymorphism for 480 species of vascular plants. They found that the mean level of genetic diversity for localized endemic species was significantly lower than the mean level of genetic diversity for species with regional or widespread distributions (Figure 10.3).

In some cases, geographical range is a poor predictor of levels of genetic variation in a species. This may reflect differences among species in the severity of past genetic bottlenecks (Bonnell and Selander, 1974; Franklin, 1980; Frankel and Soulé, 1981; Critchfield, 1984; McClenaghan and Beauchamp, 1986). It may also be due to historical changes in a species'

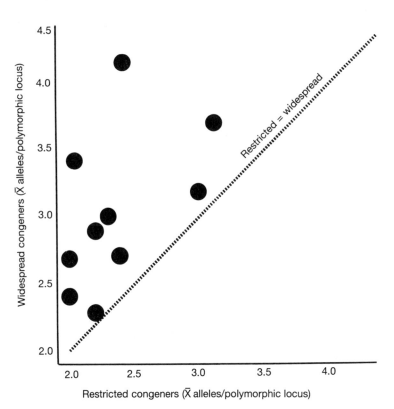

Figure 10.2 Comparison of the mean number of alleles per polymorphic locus for sets of geographically restricted species and widespread congeners in 10 vascular plant genera. If the mean values are identical for the restricted and widespread species, they will fall on the dotted line. The mean number of alleles per polymorphic locus is significantly higher for the widespread species (Wilcoxon's signed-ranks test, $P < 0.05$). (From Karron, 1987a.)

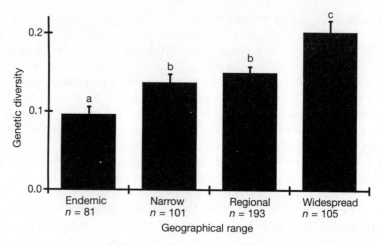

Figure 10.3 Mean levels of genetic diversity for species in each of four geographic range categories. Lower-case letters designate means that are significantly different at the 5% probability level. (From data presented in Hamrick and Godt, 1990.)

range (Harper, 1981; Kruckeberg and Rabinowitz, 1985). While some localized endemics were once widespread and have only recently declined in range, others are relatively recent in origin and have never had a large range (Stebbins and Major, 1965).

Many, but not all, sparse species have several populations encompassing a broad geographical area. We may expect widely distributed sparse species to have fairly high levels of genetic variation at the species level. However, because the effective population size of sparse species is likely to be small, much of this genetic diversity may be distributed among, rather than within, populations. Despite detailed studies of the ecology of sparse species (Rabinowitz, 1981; Rabinowitz *et al.*, 1984; Rabinowitz *et al.*, 1986), little is known about their population genetic structure. Hamrick *et al.* (1991) compared levels of polymorphism for 16 locally common and 13 locally uncommon tropical tree species on Barro Colorado Island in the Republic of Panama. They found that levels of genetic diversity were significantly higher for the locally common taxa, but that the uncommon species had moderate levels of genetic variation. As these researchers note, it would be interesting to compare levels of polymorphism on Barro Colorado Island with data from sites where the locally uncommon species occur at higher density.

10.4 BREEDING SYSTEMS OF RARE PLANT SPECIES

The extent of inbreeding in localized endemics may be influenced by both current population size and the severity of past population bottlenecks. If few mates are available for cross-fertilization (Baker, 1955, 1967; Jain,

1976) or if pollinator service is unreliable (Tepedino, 1979; Karron, 1987b; Wyatt, 1986), self-pollinating individuals may be favoured by natural selection (Lande and Schemske, 1985; Schemske and Lande, 1985; but see also Waller, 1986 and Barrett and Kohn, 1991). By contrast, if a localized endemic has only recently declined in range, or has consistently had moderate-sized populations, there may have been little selection for self-fertility.

Several localized endemics exhibit higher levels of inbreeding than do closely related widespread congeners (Karron, 1991): *Limnanthes bakeri*, a vernal pool species endemic to a valley 10 km long in Mendocino County, California, is highly auto-fertile and mostly selfing, whereas widespread *Limnanthes douglassii* primarily reproduces through cross-fertilization (Table 10.1; Kesseli and Jain, 1984, 1986). *Stephanomeria malheurensis*, only known to occur on a single hilltop south of Burns, Oregon, is also auto-fertile. By contrast, closely related widespread *Stephanomeria exigua* ssp. *coronaria* is self-incompatible (Gottlieb, 1973, 1979). Rare *Psychotria* species have a significantly higher incidence of self-compatibility and homostyly than do widespread members of this tropical genus (Chapter 11). Two geographically restricted *Astragalus* species, *A. linifolius* and *A. osterhouti*, have significantly higher levels of self-fertility than closely related widespread *A. pectinatus* (Karron, 1989).

Although many localized endemics are highly inbreeding, a few are obligate outcrossers. *Oenothera organensis*, a hawkmoth-pollinated species endemic to the Organ Mountains of New Mexico, has a well developed genetic self-incompatibility system (Emerson, 1938). Despite a low level of allozyme polymorphism due to the effects of drift (Levin *et al.*, 1979), more than 45 alleles at the gametophytic self-incompatibility locus (S-alleles) have been maintained by frequency-dependent selection (Emerson, 1939; Wright, 1939).

In some cases, severe genetic drift in small populations of self-incompatible species may lead to the loss of S-alleles (Imrie *et al.*, 1972; Byers and Meagher, 1992; Reinartz and Les, 1994). Reproductive failure could result if all individuals in a population have the same mating type.

Table 10.1 Components of self-fertilization in geographically restricted *Limnanthes bakeri* and geographically widespread *L. douglassii* (modified from Kesseli and Jain, 1986)

Component	*L. bakeri* (restricted)	*L. douglassii*[a] (widespread)
Autogamous seed-set per flower ($\bar{X}$)	2.11	0.06
Natural seed-set per flower ($\bar{X}$)	3.29	3.48
Selfing rate[b]	0.79	0.05

[a]Data for *L. douglassii* population 511.
[b]Selfing rate = (1 − multilocus outcrossing rate)

For example, *Aster furcatus* populations with several sporophytic S-alleles have high levels of seed set, whereas populations with very low levels of seed production are suspected of having three or fewer S-alleles (Les *et al.*, 1991; Reinartz and Les, 1994).

Sparse species that are pollinated by wind or generalist foragers are likely to have small effective population sizes. This may result in an increased incidence of biparental inbreeding, leading to a higher level of homozygosity in these populations. Kunin (1992, 1993) demonstrated that population density dramatically influences reproductive success in self-incompatible *Diplotaxis erucoides* and *Brassica kaber*. Thus, the scarcity of mates in low-density populations may favour the evolution of self-fertilization (Baker, 1955, 1967).

Detailed studies of mating patterns have only been completed for a small number of sparse species. Murawski and Hamrick (1991) measured outcrossing rates for nine tropical tree species on Barro Colorado Island. Mean flowering tree density for these species spanned nearly two orders of magnitude, ranging from 0.08 to 6.51 individuals per ha. Most species had very high outcrossing rates, but two of the four lowest-density species exhibited a mixture of self- and cross-fertilization. There is considerable need for additional studies of mating systems in sparse species, especially studies that compare mating patterns in closely related sparse and locally abundant taxa.

10.5 THE INFLUENCE OF POPULATION DENSITY ON MATING PATTERNS IN MONKEYFLOWER

The effects of population density on breeding systems have rarely been studied experimentally (Van Treuren *et al.*, 1993a). However, several researchers have suggested that outcrossing rates of insect-pollinated self-compatible species will be positively correlated with density (Bateman, 1956; Handel, 1983; Brown *et al.*, 1989; Murawski and Hamrick, 1991). At low density, within–plant pollinations should predominate since pollen-vectors usually fly short distances. As density increases, pollinators will tend to move among plants, resulting in a higher frequency of outcrossing.

This hypothesis has been experimentally tested (Karron *et al.*, 1995) with *Mimulus ringens* (square-stemmed monkeyflower), a bumblebee-pollinated perennial herb that is broadly distributed in central and eastern North America (Grant, 1924). Populations of this wetland species occur at a wide range of densities and typically have fewer than 50 individuals. *Mimulus ringens* is self-compatible and natural populations have a mean outcrossing rate of 0.34 (J.D. Karron, unpublished). Flowers of this taxon are similar in size and shape to those of *M. guttatus*, a species that varies dramatically in outcrossing rate among populations (Ritland and Ganders, 1987; Ritland, 1990; Dole, 1991). In addition to sexual reproduction, *M. ringens* replicates clonally by stoloniferous rhizomes (Grant, 1924).

In nature, population density is often confounded with other demographic variables such as plant size, population shape and population size (Barrett and Eckert, 1990; Kunin, 1993). In order to minimize the effects of these confounding variables, six isolated common gardens were established and two replicate arrays were planted at each of three densities (Karron *et al.*, 1995): two arrays were planted at 'high density' (0.6 m spacing between plants), two arrays were planted at 'medium density' (1.2 m spacing) and two arrays were planted at 'low density' (2.4 m spacing). This 16-fold range in density corresponds to the range typically observed in natural populations of *Mimulus ringens*. Each array had four parallel rows, with four evenly spaced plants per row.

Every plant in an array had a unique combination of homozygous genotypes at four allozyme loci (Table 10.2). This novel approach greatly increased the precision of outcrossing rate estimates because paternity could readily be determined for each sampled seed (Figure 10.4). The 16 genets were produced through a complicated programme of experimental crosses and genetic screening; details may be found in Karron *et al.* (1995). Each of the six arrays was then clonally propagated from the same set of 16 genets to minimize differences in floral and vegetative morphology among density treatments.

Movements of pollinators visiting *M. ringens* were recorded throughout the peak period of flowering in the experimental populations. The proportion of pollinator flights between plants varied significantly among populations and was greatest at high density (Figure 10.5a). Outcrossing

Table 10.2 Multilocus genotypes of 16 *Mimulus ringens* clones planted in each of six experimental gardens

Clone	Locus			
	SKDH-1	ACON-3	AAT-1	ACP-1
A	11	11	11	11
B	11	11	11	22
C	11	11	22	11
D	11	11	22	22
E	11	22	11	11
F	11	22	11	22
G	11	22	22	11
H	11	22	22	22
J	22	11	11	11
K	22	11	11	22
L	22	11	22	11
M	22	11	22	22
N	22	22	11	11
O	22	22	11	22
P	22	22	22	11
Q	22	22	22	22

Genotype of maternal parent:

ACP1	ACO3	AAT1	SKD1
1 1	2 2	2 2	1 1

Genotype of offspring produced by **selfing**:

1 1	2 2	2 2	1 1

Genotype of offspring produced by **outcrossing**:
(only 1 of 15 possible genotypes is shown)

Genotype of the only possible paternal parent that could have sired the outcrossed seed shown above:

Figure 10.4 An example of paternity assignment using experimental arrays of unique multilocus homozygous genotypes. Alleles from the maternal parent are depicted as black letters on a white background. Alleles from the paternal parent are depicted as white letters on a black background. Progeny resulting from self-fertilization will have the same multilocus genotype as the maternal parent. Progeny resulting from outcrossing will be heterozygous at one or more loci.

rates were determined from paternity data for 1920 progeny (20 randomly sampled seeds from each of 16 plants from all six populations). The effects of density on outcrossing rates were highly significant, and closely paralleled patterns of pollinator movement (Figure 10.5b; Karron *et al.*, 1995).

The positive relationship between population density and outcrossing rate is likely to have been caused by changes in patterns of pollinator behaviour. At low density, bumblebees primarily moved between flowers on a single plant. These geitonogamous (within-plant) pollinations (de Jong *et al.*, 1993) resulted in a low level of outcrossing. At the highest density, pollinators frequently moved between plants, resulting in a greater rate of outcrossing. These findings closely parallel recent experimental studies with rare *Salvia pratensis*. Van Treuren *et al.* (1993a) demonstrated that outcrossing rates of this bumblebee-pollinated mint species were positively correlated with population density, but were not significantly influenced by population size.

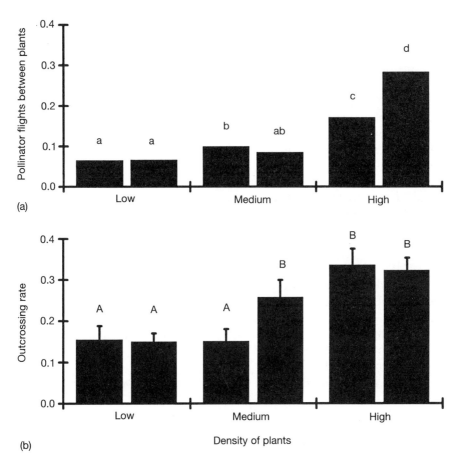

Figure 10.5 (a) Proportion of pollinator flights between plants in six experimental populations of *Mimulus ringens*. Lower-case letters designate populations with significantly different patterns of pollinator movement at the $P < 0.01$ level. (b) Outcrossing rates of pollen-fertile clones (+1 SE) in six experimental populations of *Mimulus ringens*. Upper-case letters designate populations with significantly different outcrossing rates. (Modified from Karron *et al.*, 1995.)

10.6 CONCLUSION

During the last two decades, much has been learned about the genetic structure of rare plant populations. However, many important questions await further study. Although localized endemics frequently have reduced levels of allozyme polymorphism, the extent of variation for quantitative traits that influence fitness has rarely been investigated (Ellstrand and Elam, 1993). Such studies are critically needed since quantitative variation may respond differently from nearly neutral single locus traits to the effects of small population size (Bryant *et al.*, 1986; Lande and Barrowclough, 1987; Goodnight, 1988; Willis and Orr, 1993).

The fitness consequences of inbreeding in localized endemics and sparse species are also poorly understood. Two recent studies have yielded puzzling results. Geographically restricted *Astragalus linifolius* is self-fertile and has little allozyme polymorphism, yet this species has an unusually high level of inbreeding depression (Karron, 1989). In rare *Scabiosa columbaria*, allozyme polymorphism is positively correlated with population size (Van Treuren *et al.*, 1991), yet Van Treuren *et al.* (1993b) found no relationship between inbreeding depression and population size. One explanation for these surprising results is that the small populations may have recently undergone a bottleneck, and additional generations of inbreeding and selection are necessary to purge the genetic load. These findings emphasize the need for long-term studies of the evolutionary consequences of declining population size.

ACKNOWLEDGEMENTS

I thank Nina Thumser, Rosella Tucker, Amy Hessenauer and Stephanie Schlicht for assistance in the field, greenhouse and laboratory. I am grateful to James Reinartz and Lou Nelson for designing and building experimental research gardens at the UW-Milwaukee Field Station and to Thomas Schuck for advice and assistance in propagating *M. ringens*. James Reinartz, Christophe Thebaud and an anonymous reviewer provided many helpful suggestions on an earlier draft of this chapter. Research on *Mimulus ringens* was supported by United States National Science Foundation grant DEB-9119311, and by a research award from the Graduate School of UW-Milwaukee. This is publication number 151 of the UW-Milwaukee Field Station.

REFERENCES

Baker, H.G. (1955) Self-compatibility and establishment after 'long-distance' dispersal. *Evolution*, **9**, 347–349.

Baker, H.G. (1967) Support for Baker's Law – as a rule. *Evolution*, **21**, 853–856.

Barrett, S.C.H. and Eckert, C.G. (1990) Variation and evolution of mating systems in seed plants, in *Biological Approaches and Evolutionary Trends in Plants* (ed. S. Kawano), Academic Press, London, pp. 229–254.

Barrett, S.C.H. and Kohn, J.R. (1991) Genetic and evolutionary consequences of small population size, in *Genetics and Conservation of Rare Plants* (eds D.A. Falk and K.E. Holsinger), Oxford University Press, Oxford, pp. 3–30.

Bateman, A.J. (1956) Cryptic self-incompatibility in the wall-flower: *Cheiranthus cheiri* L. *Heredity*, **10**, 257–261.

Beardmore, J.A. (1983) Extinction, survival and genetic variation, in *Genetics and Conservation* (eds C.M. Schonewald-Cox, S.M. Chambers, B. MacBryde and W.L. Thomas), Benjamin-Cummings, Menlo Park, California, pp. 125–151.

Bonnell, M.L. and Selander, R.K. (1974) Elephant seals: genetic variation and near extinction. *Science*, **184**, 908–909.

Brown, A.H.D., Burdon, J.J. and Jarosz, A.M. (1989) Isozyme analysis of plant mating systems, in *Isozymes in Plant Biology* (eds D.E. Soltis and P.S. Soltis), Dioscorides Press, Portland, Oregon, pp. 73–86.

Bryant, E.H., McCommas, S.A. and Combs, L.M. (1986) The effect of an experimental bottleneck upon quantitative genetic variation in the housefly. *Genetics*, **114**, 1191–1211.

Burdon, J.J. (1987) *Diseases and Plant Population Biology*, Cambridge University Press, Cambridge.

Byers, D.L. and Meagher, T.R. (1992) Mate availability in small populations of plant species with homomorphic sporophytic self-incompatibility. *Heredity*, **68**, 353–359.

Chapin, F.S., III and Chapin, M.C. (1981) Ecotypic differentiation of growth processes in *Carex aquatilis* along latitudinal and local gradients. *Ecology*, **62**, 1000–1009.

Clausen, J., Keck, D.D. and Hiesey, W.M. (1940) *Experimental Studies on the Nature of Species. I. The Effect of Varied Environments on Western North American Plants*, Carnegie Institute of Washington, Publ. No. 520.

Clausen, J., Keck, D.D. and Hiesey, W.M. (1948) *Experimental Studies on the Nature of Species. III. Environmental Responses of Climatic Races of* Achillea, Carnegie Institute of Washington, Publ. No. 581.

Critchfield, W.B. (1984) Impact of the Pleistocene on the genetic structure of North American conifers, in *Proceedings of the Eighth North American Biology Workshop* (ed. R.M. Lanner), Logan, Utah, pp. 70–118.

De Jong, T.J., Waser, N.M. and Klinkhamer, P.G.L. (1993) Geitonogamy: the neglected side of selfing. *Trends in Ecology and Evolution*, **8**, 321–325.

Dole, J.A. (1991) Evolution of mating systems in the *Mimulus guttatus* complex. PhD Dissertation, University of California, Davis.

Drury, W.H. (1974) Rare species. *Biological Conservation*, **6**, 162–169.

Drury, W.H. (1980) Rare species of plants. *Rhodora*, **82**, 3–48.

Ellstrand, N.C. and Elam, D.R. (1993) Population genetic consequences of small population size: implications for plant conservation. *Annual Review of Ecology and Systematics*, **24**, 217–242.

Emerson, S. (1938) The genetics of self-incompatibility of *Oenothera organensis*. *Genetics*, **23**, 190–202.

Emerson, S. (1939) A preliminary survey of the *Oenothera organensis* population. *Genetics*, **24**, 524–537.

Fiedler, P.L. and Ahouse, J.J. (1992) Hierarchies of cause: toward an understanding of rarity in vascular plant species, in *Conservation Biology: The Theory and Practice of Nature Conservation, Preservation, and Management* (eds P.L. Fiedler and S.K. Jain), Chapman & Hall, New York, pp. 23–47.

Frankel, O.H. and Soulé, M.L. (1981) *Conservation and Evolution*, Cambridge University Press, Cambridge.

Franklin, I.R. (1980) Evolutionary change in small populations, in *Conservation Biology* (ed. M.E. Soulé and B.A. Wilcox), Sinauer Associates, Sunderland, Massachusetts, pp. 135–149.

Gaston, K.J. (1994) *Rarity*, Chapman & Hall, London.

Goodnight, C.J. (1988) Epistasis and the effect of founder events on the additive genetic variance. *Evolution*, **42**, 441–454.

Gottlieb, L.D. (1973) Genetic differentiation, sympatric speciation, and the origin of a diploid species of *Stephanomeria*. *American Journal of Botany*, **60**, 545–553.

Gottlieb, L.D. (1979) The origin of phenotype in a recently evolved species, in *Topics in Plant Population Biology* (eds O.T. Solbrig, S. Jain, G.B. Johnson and P.H. Raven), Columbia University Press, New York, pp. 264–86.

Grant, A.L. (1924) A monograph of the genus *Mimulus*. *Annals of the Missouri Botanical Garden*, **11**, 99–388.

Hamrick, J.L. (1983) The distribution of genetic variation within and among natural plant populations, in *Genetics and Conservation* (eds C.M. Schonewald-Cox, S.M. Chambers, B. MacBryde and W.L. Thomas), Benjamins/Cummings, Menlo Park, California, pp. 335–348.

Hamrick, J.L. and Godt, M.J.W. (1990) Allozyme diversity in plant species, in *Plant Population Genetics, Breeding and Genetic Resources* (eds A.H.D. Brown, M.T. Clegg, A.L. Kahler and B.S. Weir), Sinauer Associates, Sunderland, Massachusetts, pp. 43–63.

Hamrick, J.L., Godt, M.J.W., Murawski, D.A. and Loveless, M.D. (1991) Correlations between species traits and allozyme diversity: implications for conservation biology, in *Genetics and Conservation of Rare Plants*, (eds D.A. Falk and K.E. Holsinger), Oxford University Press, Oxford, pp. 75–86.

Hamrick, J.L., Linhart, Y.B. and Mitton, J.B. (1979) Relationships between life history characteristics and electrophoretically detectable genetic variation in plants. *Annual Review of Ecology and Systematics*, **10**, 173–200.

Handel, S.N. 1983. Contrasting gene flow patterns and genetic subdivision in adjacent populations of *Cucumis sativus* (Cucurbitaceae). *Evolution*, **37**, 760–771.

Harper, J.L. (1981) The meanings of rarity, in *The Biological Aspects of Rare Plant Conservation* (ed. H. Synge), Wiley, Chichester, pp. 189–203.

Huenneke, L.F. (1991) Ecological implications of genetic variation in plant populations, in *Genetics and Conservation of Rare Plants* (eds D.A. Falk and K.E. Holsinger), Oxford University Press, Oxford, pp. 31–44.

Husband, B.C. and Barrett, S.C.H. (1992) Genetic drift and the maintenance of the style length polymorphism in tristylous populations of *Eichhornia paniculata* (Pontederiaceae). *Heredity*, **69**, 440–449.

Imrie, B.C., Kirkman, C.J. and Ross, D.R. (1972) Computer simulation of a sporophytic self-incompatibility breeding system. *Australian Journal of Biological Science*, **25**, 343–349.

Jain, S.K. (1976) The evolution of inbreeding in plants. *Annual Review of Ecology and Systematics*, **7**, 469–495.

Jarne, P. and Charlesworth, D. (1993) The evolution of the selfing rate in functionally hermaphrodite plants and animals. *Annual Review of Ecology and Systematics*, **24**, 441–466.

Karron, J.D. (1987a) A comparison of levels of genetic polymorphism and self-compatibility in geographically restricted and widespread plant congeners. *Evolutionary Ecology*, **1**, 47–58.

Karron, J.D. (1987b) The pollination ecology of co-occurring geographically restricted and widespread species of *Astragalus* (Fabaceae). *Biological Conservation*, **39**, 179–193.

Karron, J.D. (1989) Breeding systems and levels of inbreeding depression in geographically restricted and widespread species of *Astragalus* (Fabaceae). *American Journal of Botany*, **76**, 331–340.

Karron, J.D. (1991) Patterns of genetic variation and breeding systems in rare plant species, in *Genetics and Conservation of Rare Plants* (eds D.A. Falk and K.E. Holsinger), Oxford University Press, Oxford, pp. 87–98.

Karron, J.D., Linhart, Y.B., Chaulk, C.A. and Robertson, C.A. (1988) The genetic structure of populations of geographically restricted and widespread species of *Astragalus* (Fabaceae). *American Journal of Botany*, **75**, 1114–1119.

Karron, J.D., Thumser, N.N., Tucker, R. and Hessenauer, A.J. (1995) The influence of population density on outcrossing rates in *Mimulus ringens*. *Heredity*, **75**, 612–617.

Kesseli, R.V. and Jain, S.K. (1984) New variation and biosystematic patterns detected by allozyme and morphological comparisons in *Limnanthes* sect. Reflexae (Limnanthaceae). *Plant Systematics and Evolution*, **147**, 133–165.

Kesseli, R.V. and Jain, S.K. (1986) Breeding systems and population structure in *Limnanthes*. *Theoretical and Applied Genetics*, **71**, 292–299.

Kruckeberg, A.R. and Rabinowitz, D. (1985) Biological aspects of endemism in higher plants. *Annual Review of Ecology and Systematics*, **16**, 447–479.

Kunin, W.E. (1992) Density and reproductive success in wild populations of *Diplotaxis erucoides* (Brassicaceae). *Oecologia*, **91**, 129–133.

Kunin, W.E. (1993) Sex and the single mustard: population density and pollinator behavior effects on seed-set. *Ecology*, **74**, 2145–2160.

Kunin, W.E. and Gaston, K.J. (1993) The biology of rarity: patterns, causes, and consequences. *Trends in Ecology and Evolution*, **8**, 298–301.

Lacy, R.C. (1992) The effects of inbreeding on isolated populations: are minimum viable population sizes predictable?, in *Conservation Biology: The Theory and Practice of Nature Conservation, Preservation, and Management*, (eds P.L. Fiedler and S.K. Jain), Chapman & Hall, New York, pp. 277–296.

Lande, R. and Barrowclough, G.F. (1987) Effective population size, genetic variation, and their use in population management, in *Viable Populations for Conservation*, (ed. M.E. Soulé), Cambridge University Press, pp. 87–123.

Lande, R. and Schemske, D.W. (1985) The evolution of self-fertilization and inbreeding depression in plants. I. Genetic models. *Evolution*, **39**, 24–40.

Les, D.H., Reinartz, J.A. and Esselman, E.J. (1991) Genetic consequences of rarity in *Aster furcatus*, a threatened, self-incompatible plant. *Evolution*, **45**, 1641–1650.

Levin, D.A., Ritter, K. and Ellstrand, N.C. (1979) Protein polymorphism in the narrow endemic *Oenothera organensis*. *Evolution*, **33**, 534–542.

Linhart, Y.B. (1974) Intra-population differentiation in annual plants. I. *Veronica peregrina* L. raised under non-competitive conditions. *Evolution*, **28**, 232–243.

Loveless, M.D. and Hamrick, J.L. (1984) Ecological determinants of genetic structure in plant populations. *Annual Review of Ecology and Systematics*, **15**, 65–95.

McClenaghan, L.R. and Beauchamp, A.C. (1986) Low genic differentiation among isolated populations of the California fan palm (*Washingtonia filifera*). *Evolution*, **40**, 315–322.

McGraw, J.B. and Antonovics, J. (1983) Experimental ecology of *Dryas octopetala* ecotypes. I. Ecotypic differentiation and life-cycle stages of selection. *Journal of Ecology*, **71**, 879–897.

McNeilly, T. and Bradshaw, A.D. (1968) Evolutionary processes in populations of copper tolerant *Agrostis tenuis* Sibth. *Evolution*, **22**, 108–118.

Menges, E.S. (1991) The application of minimum viable population theory to plants, in *Genetics and Conservation of Rare Plants* (eds D.A. Falk and K.E. Holsinger), Oxford University Press, Oxford, pp. 45–61.

Murawski, D.A. and Hamrick, J.L. (1991) The effect of the density of flowering individuals on the mating system of nine tropical tree species. *Heredity*, **67**, 167–174.

Nei, M., Maruyama, T. and Chakraborty, R. (1975) The bottleneck effect and genetic variability in populations. *Evolution*, **29**, 1–10.

Rabinowitz, D. (1981) Seven forms of rarity, in *The Biological Aspects of Rare Plant Conservation* (ed. H. Synge), John Wiley, New York, pp. 205–217.

Rabinowitz, D., Cairns, S. and Dillon, T. (1986) Seven forms of rarity and their frequency in the flora of the British Isles, in *Conservation Biology: the Science of Scarcity and Diversity* (ed. M.E. Soulé) Sinauer Associates, Sunderland, Massachusetts, pp. 182–204.

Rabinowitz, D., Rapp, J.K. and Dixon, P.M. (1984) Competitive abilities of sparse grass species: means of persistence or cause of abundance. *Ecology*, **65**, 1144–54.

Reinartz, J.A. and Les, D.H. (1994) Bottleneck-induced dissolution of self-incompatibility and breeding system consequences in *Aster furcatus* (Asteraceae). *American Journal of Botany*, **81**, 446–455.

Ritland, K. (1990) Inferences about inbreeding depression based on changes of the inbreeding coefficient. *Evolution*, **44**, 1230–1241.

Ritland, K. and Ganders, F.R. (1987) Covariation of selfing rates with parental gene fixation within populations of *Mimulus guttatus*. *Evolution*, **41**, 760–771.

Schemske, D.W. and Lande, R. (1985) The evolution of self-fertilization and inbreeding depression in plants. II. Empirical observations. *Evolution*, **39**, 41–52.

Schmidt, K.P. and Levin, D.A. (1985) The comparative demography of reciprocally sown populations of *Phlox drummondii*. I. Survivorships, fecundities, and finite rates of increase. *Evolution*, **39**, 396–404.

Silander, J.A. (1979) Microevolution and clone structure in *Spartina patens*. *Science*, **203**, 658–660.

Stebbins, G.L. and Major, J. (1965) Endemism and speciation in the California flora. *Ecological Monographs*, **35**, 1–35.

Tepedino, V.J. (1979) The importance of bees and other insect pollinators in maintaining floral species composition. *Great Basin Naturalist Memoirs*, **3**, 139–150.

Van Dijk, H. (1989) Genetic variability in *Plantago* species in relation to their ecology. 4. Ecotypic differentiation in *P. major*. *Theoretical and Applied Genetics*, **77**, 749–759.

Van Treuren, R., Bijlsma, R., Ouborg, N.J. and Van Delden, W. (1993a) The effects of population size and plant density on outcrossing rates in locally endangered *Salvia pratensis*. *Evolution*, **47**, 1094–1104.

Van Treuren, R., Bijlsma, R., Ouborg, N.J. and Van Delden, W. (1993b) The significance of genetic erosion in the process of extinction. IV. Inbreeding depression and heterosis effects caused by selfing and outcrossing in *Scabiosa columbaria*. *Evolution*, **47**, 1669–1680.

Van Treuren, R., Bijlsma, R., Van Delden, W. and Ouborg, N.J. (1991) The significance of genetic erosion in the process of extinction. I. Genetic differentiation in *Salvia pratensis* and *Scabiosa columbaria* in relation to population size. *Heredity*, **66**, 181–189.

Waller, D.M. (1986) Is there disruptive selection for self-fertilization? *American Naturalist*, **128**, 421–426.

Willis, J.H. and Orr, H.A. (1993) Increased heritable variation following population bottlenecks: the role of dominance. *Evolution*, **47**, 949–957.

Wright, S. (1939) The distribution of self-sterility alleles in populations. *Genetics*, **24**, 538–552.

Wright, S. (1969) *Evolution and the Genetics of Populations. Volume 2: The Theory of Gene Frequencies*, University of Chicago Press, Chicago.

Wright, S. (1977) *Evolution and the Genetics of Populations. Volume 3: Experimental Results and Evolutionary Deductions*, University of Chicago Press, Chicago.

Wyatt, R. (1986) Ecology and evolution of self-pollination in *Arenaria uniflora* (Caryophyllaceae). *Journal of Ecology*, **74**, 403–418.

11 Evolved consequences of rarity

Gordon H. Orians

11.1 INTRODUCTION

Rarity is common and widespread. As early as 1859 Charles Darwin wrote that 'rarity is the attribute of a vast number of species in all classes, in all countries'. All species are rare somewhere, and most species are rare toward the periphery of their ranges (Hanski, 1982; Brown, 1984; Hanski *et al.*, 1993). The majority of species comprising local biotas are rare (Preston, 1948; Hubbell, 1979). Indeed, the rarity of most species gave rise to a major school of ecological thought, which asserted that populations were typically regulated by density-independent factors acting at low population densities rather than by density-dependent factors, such as competition, whose effects should be most strongly felt at high population densities (Andrewartha and Birch, 1954).

Many species that are rare in some places are common elsewhere, but some species are rare everywhere. Newly formed species generated by polyploidy or founder populations are rare. Species nearing extinction or at advanced stages in a taxon cycle are rare (Wilson, 1961; Ricklefs and Cox, 1972). Species high on the trophic ladder and specialists on rare resources are typically rare. Many species that were formerly common are being made rare by human activities such as over-exploitation and habitat destruction. Concern over the large numbers of species whose survival is being threatened by human activities has stimulated considerable research on the causes of rarity and means of prolonging the survival of rare species (Noss and Cooperrider, 1994). Such activities are of considerable theoretical and practical importance, but the consequences, as opposed to the causes, of rarity have received relatively little attention.

How rarity might influence and be influenced by evolutionary processes has concerned biologists ever since Darwin suggested that abundant species were more likely than rare species to be the sources of evolutionary novelties. Nonetheless, despite considerable theoretical and empirical

The Biology of Rarity. Edited by William E. Kunin and Kevin J. Gaston.
Published in 1997 by Chapman & Hall, London. ISBN 0 412 63380 9.

attention to the matter, evolutionary biologists today hold widely varying views on the role of rarity in evolution (Chapter 12). This is, perhaps, inevitable given that the outcomes of the interaction between rarity and evolution, both of which are complex entities, are certain to be highly varied.

This chapter explores interactions between rarity and evolution by focusing on those traits that may be especially favoured by natural selection in species that are rare for evolutionarily significant time spans. More specifically, it focuses on characteristics of the reproductive biology of plants that might be sensitive to population density and range size. It does not discuss many interesting issues such as how rarity influences the rate of evolution, how to estimate relevant population sizes for defining rarity, or the role of rarity in speciation.

Rarity can usefully be measured using three criteria: geographical range (narrow), habitat tolerance (narrow) and local population density (low). Seven combinations of these traits constitute reasonable categories of rarity (Rabinowitz, 1981). In addition, populations of a species may be patchily distributed even in appropriate habitats, adding another dimension to the Rabinowitz typology of rarity.

11.2 WHY SHOULD RARE SPECIES DIFFER FROM COMMON ONES?

There are many reasons why rare species might evolve traits that, on average, differ from those of their more common close relatives. Compared with those of common species, individuals of species that have low population densities may have more difficulty in locating mates, are more likely to have heterospecific neighbours, and are more likely to be fed upon by generalist predators and parasites. All of these factors alter the dynamics of rare populations in ways that may favour the evolution of traits that enhance success under conditions of low population density.

To investigate the consequences of rarity, a number of investigators have compared traits of rare and common species. Rabinowitz and her co-workers predicted, and in experimental studies confirmed, that species of grasses having low local population densities differed from common species in their dispersal characteristics (Rabinowitz and Rapp, 1981), competitive abilities (Rabinowitz et al., 1984), palatability (Landa and Rabinowitz, 1983) and variability of reproductive success (Rabinowitz et al., 1989). Some workers report differences between rare and common plant species in growth form, floral morphology and mode of pollination (Harper, 1979), but other workers fail to find such differences (Fiedler, 1987).

Several factors could lead to variable outcomes in surveys of the traits of rare and common species. Comparisons made in the absence of a theoretical predictive framework are likely to incorporate large amounts of phylogenetic 'noise' and irrelevant traits into the comparisons. Rare species in different categories of Rabinowitz's classification would not necessarily be expected to share similar suites of traits. Newly rare species are likely to differ in many characteristics from old-time rare ones.

However, even if consistent differences between rare and common species are identified, natural selection need not be responsible for them. Organisms with certain characteristics may be disproportionately likely to become rare or extinct (Terborgh and Winter, 1980). Thus, over-representation of rare species among self-compatible plants could result from selective extinction of self-incompatible species or increased probability of local speciation (and thus initial rarity) among selfing taxa, as well as from selection favouring self-compatibility in rare populations. Similarly, newly formed rare species in lineages that are speciating via polyploidy may share many traits with their common congeners, none of which have been moulded by natural selection to favour success under conditions of restricted ranges.

Because of the difficulties in interpreting differences between rare and common species identified in comparative studies, studies of intraspecific variation in traits associated with rarity are an important supplement to interspecific comparisons. Because all populations of a species share a recent common ancestor, ancestral and derived states of intraspecifically varying characters are readily determined. A potential weakness with intraspecific comparisons is that the populations being compared may not have been rare or common long enough for evolutionary changes to have happened. Nevertheless, consistent differences in characteristics of rare and common populations of the same species constitute strong evidence for evolutionary changes driven by natural selection.

11.3 PREDICTING DIFFERENCES BETWEEN RARE AND COMMON SPECIES

To distinguish traits of rare species that have evolved under the influence of natural selection from those due to other, nonadapative causes, probable population-dynamic consequences of rarity need to be specified and used as a basis for predicting phenotypic changes that might enhance survival and reproductive success under conditions of rarity. Among the three basic components of rarity, local population density lends itself more readily to clear predictions than does habitat specialization or total geographical range. However, because widespread species are generally found at higher population densities, particularly in central parts of their ranges, than species with restricted ranges (Hanski, 1982; Brown, 1984), geographical range can, with caution, be used as a surrogate for local population density when data about the latter are lacking. This chapter concentrates on the evolutionary consequences of low local population densities.

11.3.1 Identity of neighbours

The neighbours of individuals of rare species are more likely to be members of other species than are the neighbours of individuals of common species. Therefore, natural selection may favour stronger interspecific competitive ability in rare species, but stronger intraspecific competitive ability in common species (Rabinowitz et al., 1984).

11.3.2 Enemies

By virtue of being locally or sparsely distributed, rare plants may escape in space and time from herbivores and pathogens (Feeny, 1976). In addition, many parts of plants are present only periodically. Leaves of most plants live for many months, but flowers of many species are present for only a few days each year, and no flowers may be produced locally during some years. Seed production is strongly correlated temporally with flowering, but because of their longevity, seeds typically are present for much longer periods than flowers.

Because monophagy or oligophagy are less likely to evolve among consumers of rare species than among consumers of common species, individuals of rare species are less likely than common ones to be fed upon by specialized herbivores. If so, their defences should evolve to be more strongly directed toward generalist predators, whereas more common species may be defended more strongly against specialist predators. Some evidence for this pattern is available for the chemical defences of a variety of plant taxa (Cates and Orians, 1975; Feeny, 1976; Rhoades and Cates, 1976; Berenbaum, 1981a,b) but the pattern is not universal (Futuyma, 1976). The type and level of plant chemical defences also depends on the environment in which the plant grows (Coley et al., 1985).

Although flowers typically are available for only short time spans, they are fed upon by many animals, probably because they are generally less well defended than leaves (Chew, 1975; Bazzaz et al., 1987). Flowers are a major or sole food source for some insect species in nine orders (Burgess, 1991). Floral traits that increase attractiveness to pollinators – large conspicuous flowers, large rewards, long-lived flowers – are likely to increase their suitability as food for florivores. Specialized florivores are most likely to evolve on species in which individual plants flower for extended periods and movement between flowers on a plant is easy. Because rare species are less likely than common species to have specialized florivores, regardless of the traits of their flowers, counter-selection against large long-lived flowers that provide large rewards is probably weaker in rare species than counter-selection against such traits in common species. If so, large long-lived flowers should evolve more easily in rare than in common species, unless selection for increased floral rewards overcomes selection for traits that reduce florivory.

11.3.3 Relationships with mutualists

Rare and common species should differ in their relationships with mutualists, particularly interactions among plants and their pollinators and seed dispersers. The individuals of a plant species living at low local population densities are less likely to be visited by specialized pollinators than are individuals of an abundant species, and individuals of generalized pollinators are less likely to 'major' on rare species than on common species. Therefore, pollinators of rare plants are likely to visit a number of plants of other species before their next visit to another conspecific plant, with major consequences for the efficiency of pollen transfer. In particular, isolated individuals should have lower reproductive success than individuals closer to conspecifics, a pattern that has been found in a number of species (see summary in Levin, 1995).

11.3.4 Finding mates

Motile organisms, even if they are rare, can find mates by following airborne chemical gradients or assembling at particular topographical features (Thornhill and Alcock, 1983). Sessile organisms, on the other hand, depend either on the movement of the medium to disperse their sexual propagules or they use the services of some vector. The gametes of many sessile marine animals are carried by water currents to receptive sexual partners. The pollen of some plants is dispersed by the physical medium, typically the wind, but the pollen of most species is carried by animals. Because plants are rooted to one spot, their distances to nearest neighbours are fixed for long periods, although distances may change as a result of death, germination and growth of individuals. Traits that increase the probability that a pollinator leaving an individual still retains and deposits pollen it has removed from the plant when it visits the next conspecific individual should be more strongly favoured in rare species than in common ones. Alternatively, self-compatibility may be favoured in species growing at low population densities or in outlying individuals of common species (Smyth and Hamrick, 1984; Sih and Baltus, 1987; Jennersten, 1988; Murawski et al.; 1990; Oostermeijer et al., 1992), even though their inbred progeny may have a higher incidence of genetic defects (Squillace and Krause, 1963; Bannister, 1965).

11.4 RARITY AND THE REPRODUCTIVE SUCCESS OF FLOWERING PLANTS

Because of the widespread dependence of plants on animal vectors of their pollen, and the extensive information available about the structures of flowers and inflorescences and temporal patterns of flowering, angiosperms are an excellent group with which to explore the consequences of rarity for the evolution of reproductive characteristics. The remainder of this chapter concentrates on how evolution of the reproductive traits of flowering plants might be influenced by their population densities and their interactions with pollinators. Comparable relationships do not exist between plants and the dispersers of their fruits because there is no 'payment upon delivery' component to seed dispersal and no fixed target to which seeds must be dispersed (Wheelwright and Orians, 1982).

A plant's floral morphology and rewards affect the number and mix of pollinator visits it receives. These visits, together with the plant's breeding system, determine the influence of those visits on the number and genetic make-up of its progeny. The reproductive success of plants with low population densities may be strongly influenced by pollinator behaviour. The nature and severity of those influences varies according to the pollinator mix, pollinator abundance, and the kinds and abundances of other plants concurrently offering rewards to flower visitors (Feinsinger *et al.*, 1986). Nonetheless, the influences of pollinator behaviour on the reproductive success of outcrossing rare plants fall into two major categories.

For an outcrossing plant with hermaphroditic flowers, reproductive success is strongly influenced by its ability to attract visitors that carry and deposit compatible pollen on its receptive stigma. The reproductive success of all animal-pollinated plants is influenced by the state of arriving pollinators, but influences are probably stronger among rare species where arriving pollinators are more likely to have visited heterospecific flowers since their last visit to a conspecific flower. The second category, which is connected through pollinator behaviour to the first, is the probability that a visitor will carry some pollen to receptive conspecific stigmas. The rarer the plant, the greater the number of heterospecific plants a pollinator is likely to visit before it encounters another conspecific plant and, thus, the greater the loss of pollen it is carrying. Among dioecious species, the dynamics are different because pollen flows unidirectionally. The success of staminate flowers is influenced primarily by the behaviour of a pollinator after it leaves a plant, whereas the success of pistillate flowers is determined primarily by prior pollinator behaviour.

Pollinators are foragers whose behaviour typically functions to maximize the rates at which they harvest floral resources (Real, 1981; Wolf and Hainsworth, 1983; Cartar and Dill, 1990; Gass and Roberts, 1992). Their contributions to plant reproductive success are incidental byproducts of their resource-harvesting activities. Therefore, traits that might improve the reproductive success of plants can evolve only if the behavioural changes that they favour also improve the fitnesses of the flower visitors. This fact constrains the array of floral traits that can evolve, but these limitations on the evolution of plant–pollinator relationships actually make it easier to formulate evolutionarily plausible predictions.

Reproductive traits of animal-pollinated plants that may evolve as a consequence of rarity can be grouped into five categories:

1. Traits that increase the number of pollinator visits.
2. Traits that increase the efficiency of pollen transfer.
3. Traits that increase the competitiveness of pollen grains.
4. Traits that increase the proportion of conspecifics that are suitable pollen recipients and donators.
5. Traits that reduce dependence on animal vectors.

11.4.1 Traits that may increase the number of pollinator visits

The frequency with which pollinators visit a plant may be increased if the plant offers large rewards, if its rewards are easily accessible to a variety of visitors, if its flowers are showy or are mimics of flowers that offer high rewards, or if its flowers are mimics of other attractive objects, such as receptive females of the pollinator's species. The probability that a flower is visited should be positively correlated with its longevity. Other things being equal, compared with their common relatives, rare plants should:

- have more conspicuous flowers;
- have more flowers open per day;
- offer larger and more readily accessible floral rewards;
- have longer-lived flowers, provided that visitation rates are low;
- have facultatively long-lived flowers that maintain their receptivity until they are pollinated.

The prediction that rare plants should have long-lived flowers is supported by the fact that, whereas about 85% of tropical plant species have flowers that last less than one day, the flowers of most orchids last a week or longer (Primack, 1985). Orchids of most species have very low densities of flowering individuals, both because they exist at low densities and because flowering by individual plants is very infrequent. However, whether rare orchids have, on average, longer-lived flowers than more common orchids is not known.

To test the hypothesis that flower longevity should be inversely related to population density, data were gathered on 21 species of rare and common *Psychotria* (Rubiaceae) in forest understorey in Costa Rica and Panama (G.H. Orians, unpublished data). The hypothesis was rejected because all species had flowers that lasted less than one day. Visitation rates to flowers of both rare and common species were so high that no pollen was left on the stamens of individuals of any species by early afternoon. If visitation rates by pollinators are not strongly correlated with abundance, selection for long-lived flowers may be weak or non-existent even among rare species. Therefore, comparative studies of flower longevity need to be accompanied by studies of the identity and behaviour of flower visitors.

11.4.2 Traits that may increase pollen transfer efficiency

Pollen transfer efficiency, defined as the proportion of pollen that is carried to and deposited on receptive conspecific stigmas, is influenced by:

- pollinator constancy – the probability that the next plant visited by the pollinator belongs to the same species;
- the probability that pollen is brushed off the pollinator when it visits heterospecific plants;
- the probability that the pollen is deposited on the stigma of a conspecific plant when the pollinator visits it.

(a) Pollinator constancy

The probability that a pollinator visits primarily or only plants of a single species during specific foraging bouts should be positively correlated with:

- the quantity of reward offered by the plant;
- the distinctiveness and showiness of its flowers;
- the degree to which access to the reward by other species of pollinators is restricted;
- the density of other plants concurrently offering rewards;
- the effective local density of flowering individuals of the focal plant species.

Effective local density can be increased by highly synchronized timing of reward production (diurnally and seasonally). Producing rewards at times when few other plants are blooming may result in greater pollinator constancy than if a species produces rewards simultaneously with many other species.

Among predictions that follow from these arguments are that flowering by individuals of rare species should be highly synchronized seasonally; and that flowers of rare species should offer rewards primarily at times of the day or night when few other local plants are offering rewards. There do not seem to be any good comparative tests of these predictions. On Barro Colorado Island, Panama, most species of *Psychotria* – and most other understorey shrubs as well – initiate flowering shortly after the beginning of the rains in May (Croat, 1978). The only local *Psychotria* with highly asynchronous flowering is *P. marginata*, a very common species on BCI that flowers during the dry season (Croat, 1978; Hamilton, 1990). Contrary to the prediction about synchrony of flowering, during 1991 the overall length of flowering periods of rare and common species of *Psychotria* on BCI was the same, about 45 days (G.H. Orians, unpublished data).

(b) Improved mechanics of pollen transfer

The reproductive contribution of a plant is increased if its pollen grains are still on a pollinator when it arrives at the next conspecific plant and if the pollen is deposited on the stigmas of that plant rather than being carried on to other plants. The probability that a pollen grain leaving a plant is deposited on a receptive conspecific stigma may be increased by accurate placement of pollen on the body of the pollinator, flower structures that increase the likelihood that the pollen adhering to the pollinator contacts the stigma of conspecific flowers but avoids the stigmas of heterospecific flowers, and by high intraspecific pollen–stigma affinity. If pollinator visits are few, selection may favour packaging pollen into larger units such that one or a few visits remove all of them (Nilsson, 1992). Therefore, compared with their more common congeners, rare plants should:

- have flowers with structures that result in placement of pollen on the bodies of visitors in places that differ from those where pollen of common species is deposited;
- more often have triggering mechanisms that propel pollen to specific places on the bodies of pollinators;
- have flowers that deposit pollen in places on the pollinator's body that are likely to brush against conspecific, but not against heterospecific, stigmas;
- have stigmas with features that increase their affinity to conspecific pollen while decreasing their affinity to heterospecific pollen (affinity caused by physical and chemical means should coevolve with features of the pollen grains);
- have pollen grains that are packed as tetrads, viscin threads or pollinia;
- have traits that are especially attractive to long-distance pollinators – birds, bats, euglossine bees – that are more likely to carry pollen across longer interplant distances.

Precise placement of pollen should be enhanced if the structure of a flower permits only one entrance route for pollinators; that is, if it is strongly zygomorphic. Zygomorphy is known to increase the precision of placement of pollen on the bodies of pollinators (Macior, 1978), and the complex structure of orchid flowers results in unique and precise placement of pollinia on the bodies of pollinators (Dodson, 1962). However, the degree to which such placement increases the efficiency of pollen transfer and whether the precision of pollen placement is positively correlated with the rarity of species has not been investigated for any taxon. Some suggestive information is provided by Hamilton's survey of Mesoamerican *Psychotria*, which shows that intermorph floral asymmetry is more prevalent among rare than among common distylous species of *Psychotria*. Intermorph asymmetry is a condition in which the amount of anther–stigma separation differs between pins and thrums. Only three of 14 distylous species with large geographical ranges had asymmetrical flowers, whereas four of nine species with medium ranges and four of seven species with small ranges had asymmetrical flowers (Hamilton, 1990). These interesting results need to be supplemented by detailed studies of visitor behaviour and pollen placement to determine whether floral asymmetry results in more precise placement of pollen on bodies of visitors.

(c) Competitiveness of pollen

The rarer a plant, the greater the proportion of pollen landing on receptive stigmas of its flowers that are likely to be from other species. Mechanisms that inhibit germination and growth of inappropriate pollen should be well developed among rare plants. Even so, late-arriving pollen grains are potentially at a disadvantage that may be partly offset if they can germinate quickly and grow rapidly down the style (Mulcahy, 1974; Mulcahy and Mulcahy, 1975; Snow and Spira, 1991). The ancestral condition in seed plants is to have single-aperturate pollen grains, but shifts from one to three or four apertures have occurred a number of times during angiosperm evolution, with the result that the majority of extant species produce pollen grains with multiple apertures (Donoghue and Doyle, 1989). In *Viola diversifolia*, pollen grains with four apertures germinate faster than those with three, giving them a potential advantage, but four-aperturate pollen do not survive as long as three-aperturate pollen (Dajoz *et al.*, 1991).

If pollen longevity generally is inversely proportional to aperture number, it would be difficult to predict how natural selection acts on the morphology of pollen grains of rare plants. Faster germination should, in general, be favoured, but the time interval between removal of a pollen grain from an anther to its deposition on a receptive stigma may, on average, be longer for pollen grains of rare species. The selective balance of this trade-off may depend on environmental conditions. If desiccation in dry environments is the main cause of shorter life spans of multi-aperturate pollen, then a high aperture number might be favoured in rare plants of humid environments but not in rare plants of more arid regions. Comparative studies of pollen morphology are needed to elucidate these possibilities.

11.4.3 Traits that increase the proportion of conspecifics that are potential pollen donators and recipients

For most obligatorily outcrossing plants, all flowering individuals are potential receivers and donators of pollen to one another. Some plants, however, have traits that prevent them from receiving pollen from and donating pollen to some other individuals. One such system is heterostyly. In a typical heterostylous species, about half the individuals bear flowers with long styles and short stamens, the 'pin' morph; the other half bear flowers with short styles and long stamens, the 'thrum' morph. Heterostyly is usually associated with a physiological self-incompatibility system in which only intermorph pollinations are successful. Thus, individuals are not only self-incompatible, they are also cross-incompatible with roughly half the population. As a result, the effective density of reproductive individuals is about half the value it would have if all individuals could cross-fertilize. Therefore, heterostyly should make it especially difficult for outcrossing plants to set seed when they grow under low population densities.

The prediction that deviations from heterostyly are more likely in rare than in common species can be tested with the family Rubiaceae (Richards, 1986), in particular the genus *Psychotria*, which contains the greatest number of heterostylous species of any angiosperm genus (Baker, 1958). Heterostyly may have evolved more than once in the Rubiaceae (Anderson, 1973; Barrett and Richards, 1990), but because heterostyly is taxonomically widespread and numerically predominant in the tribe Psychotrieae, deviations from heterostyly in the genus are probably derived rather than ancestral (Hamilton, 1990).

Most species of *Psychotria* produce small short-lived flowers that are visited primarily by small generalist pollinators that fly only short distances. Therefore, individuals of *Psychotria* should experience sharply declining reproductive success as distances to their nearest conspecific neighbours increase. If so, selection may favour self-compatibility or intramorph compatibility in rare species. Because the genetics involved in the breakdown of heterostyly are relatively simple (Richards, 1986; Barrett, 1989), one of the derived floral states could evolve readily under conditions of low population density, either because of simple distance effects or because genetic drift results in strong numerical dominance of one floral morph in local populations. Indeed, intramorph compatibility has been reported in some species of *Psychotria* (Baker, 1958; Bawa and Beach, 1983) and other heterostylous genera (Ornduff, 1966; Martin, 1967).

Rarity is apparently positively associated with deviations from heterostyly in *Psychotria* at both geographical range and local population density scales. From a detailed analysis of herbarium specimens, Hamilton (1989a,b,c) determined the ranges of 66 species of *Psychotria* of the sub-genus *Psychotria* in Mesoamerica. If at least five individuals of a species all had flowers of one morph, he judged the species to be monomorphic. Enough specimens existed for him to determine the floral morphology of 50 of the 66 species (Hamilton, 1990). His analysis shows that monomorphy, which includes pin-monomorphy, thrum-monomorphy and homostyly, is disproportionately represented among species with very small ranges (Table 11.1).

Floral morphology for 22 species of *Psychotria* on Barro Colorado Island, Panama, was determined in 1991. Crude estimates of the abundances of species were made using information in Croat (1978), via conversations with various investigators at BCI, and by extensively walking trails on the island. All 10 species judged to be common are distylous, as are both 'uncommon' species. Of the 10 rare species, however, only three are distylous (Table 11.2). Five of the monomorphic species are pin-monomorphic and two are homostylous. The predominance of pin-monomorphy accords with data gathered by Hamilton (1990) showing that

Table 11.1 Range size and floral morphology of Mesoamerican *Psychotria* (data from Hamilton, 1990)

Flower type	Range size (no. of 1° × 1° map squares)			
	Large > 15	Medium 7–15	Small 2–6	Very small 1
Heterostylous	14	8	15	4
Pin-monomorphic	0	1	1	1
Thrum-monomorphic	0	0	0	1
Homostylous	0	1	0	4
Unknown	0	0	2	14
Total	14	10	18	24
Proportion derived (of known taxa)	0.00	0.20	0.06	0.60

Table 11.2 Local abundance and *Psychotria* flower structure, Barro Colorado Island (BCI), Panama

Abundance on BCI	Flower type		Breeding system
	Heterostylous	Monomorphic	
Common	P. acuminata		NA
	P. chagrensis		InterMC
	P. deflexa		NA
	P. emetica		InterMC
	P. furcata		NA
	P. horizontalis		InterMC
	P. ipecacuanha		NA
	P. limonensis		InterMC
	P. marginata		InterMC
	P. pittieri		NA
Uncommon	P. capitata		NA
	P. grandis		NA
Rare	P. pubescens	P. brachybotria (pin)	NA
	P. psychotriifolia	P. brachiata (homo)	Autog
	P. tomentosa	+P. carthagenensis (pin)	NA
		+P. graciliflora (pin)	NA
		+P. micrantha (pin)	SelfC
		P. racemosa (homo)	NA
		P. tenuifolia	NA

InterMC = intermorph compatible; SelfC = self-compatible; Autog = autogamous; homo = homostylous; pin = pin monomorphy; thrum = thrum monomorphy; + = reported to be heterostylous elsewhere; NA = not available.

fruit set was higher among pins than among thrums in both *P. marginata* and *P. horizontalis* at BCI. These data support the hypothesis of Beach and Bawa (1980) that short-tongued pollinators, the most common visitors at flowers of most *Psychotria* species, effect fruit sets that are biased in favour of pins. Thus, low density populations may be more likely to lose the thrum morph than the pin morph. Interestingly, *P. psychotriifolia* appears to be pin-monomorphic at La Selva, where it is rare. The three monomorphic rare species of *Psychotria* on BCI preceded by a plus sign on Table 11.2 have been reported to be distylous elsewhere, but information on local population densities in those areas is lacking.

In general, patterns of floral morphology in the genus *Psychotria* support the prediction that deviations from heterostyly have evolved preferentially among rare species. The most likely interpretation of the patterns is that reduced seed-set by individuals of rare species has favoured monomorphy and the evolution of intramorph compatibility, but measures of correlations between isolation and seed-set in outcrossing heterostylous species have not yet been made.

11.4.4 Reduced dependency on pollination

If individuals of a rare species experience low reproductive success because of their degree of isolation, natural selection may favour self-compatibility, facultative autogamy, and increased use of vegetative reproduction. Therefore, compared with their more common congeners, rare plants should be more likely to:

● be self-compatible, facultatively autogamous, and be able to reproduce vegetatively;
● have an ability to resorb unharvested nectar.

There seems to be no comparative data on ability of plants to resorb nectar, though increased incidence of self-compatibility has been reported in rare species in some plant taxa (Mehrhoff, 1983; Fiedler, 1987; Kunin, 1991) but not all (Karron, 1987).

The five common species of *Psychotria* species on BCI for which experimental pollination data are available are strictly intermorph compatible, whereas among the rare species, *P. brachiata* is autogamous, *P. micrantha* is self-compatible and *P. tenuifolia* is probably autogamous. Individual *P. tenuifolia* plants with 18, eight and one open flowers were watched for 2–4½ hours each, during which no visits by pollinators were seen (personal observation).

11.4.5 Multifunctional traits

Some floral traits may serve several functions. For example, large rewards favour both increased visitation rates and greater pollinator constancy. In other cases, however, opposite traits may be favoured depending on the relative strengths of different effects. Thus, open access to rewards may favour high visitation rates but reduce the efficiency of pollen transfer.

It is evident that traits that contribute strongly to reproductive success in rare plants have the potential to increase reproductive success of all plants, regardless of their local abundance. Nonetheless, most of these traits have associated energy costs. Longer-lived flowers are more expensive to construct, with the result that fewer can be produced for a given amount of energy (Primack, 1985; Ashman and Schoen, 1994). Rewards are costly, and construction costs of zygomorphic flowers are likely to be higher than those of actinomorphic flowers. Other features are likely to provide little benefit to a common plant because many pollinators may major on common plants regardless of their flower structure or reward size. Therefore, the traits predicted to be disproportionately represented among rare species should be ones that have a greater potential for increasing the fitness of rare than of common plants.

11.5 DISCUSSION

Because plants live in varied environments and interact with highly variable suites of pollinators, florivores and herbivores, uniform evolutionary responses of reproductive structures to low population densities are not to be expected. Flower size and longevity are evidently conservative among species of *Psychotria*, but these same traits are evolutionarily labile in other taxa. For example, flower longevity, size and shape are notoriously variable among orchids, the archetypal family of rare plants. In Australia, the author is gathering data on the floral biology of *Eremophila* species (Myoporaceae), a genus with remarkably variable flowers. Because great variation among plant lineages will doubtless prove to be the norm, cross-lineage comparisons may be relatively uninformative until many within-lineage comparisons have been made.

A floral morph may be more readily lost than regained in heterostylous species (Richards, 1986; Barrett and Richards, 1990). Nonetheless, heterostyly remains prevalent among Mesoamerican *Psychotria* and monomorphy appears to be restricted primarily to species that either have small ranges or are locally rare. This pattern could result if extinction rates of local monomorphic populations are high but recolonization rates from high density heterostylous populations also are high. Such dynamics would be difficult, if not impossible, to observe directly, but more extensive data on intraspecific patterns of variability in floral morphology would shed considerable light on the problem.

Data on pollen carryover are scarce and are certain to be difficult to gather, but much can probably be learned from a comparative analysis of stigma and pollen structures and tests of pollen–stigma affinities. Plant species vary greatly in their stigma and pollen morphology and pollen physiology, the significance of which is still obscure (Erdtman, 1966; McNeill and Crompton, 1978; Hoekstra, 1983). Wind-pollinated plants have structurally simpler pollen grains than animal-pollinated plants, but correlations between the structure of stigmas and pollen and their influence on stigma–pollen affinities remain to be investigated.

Compared with plant–pollinator interactions, relationships between plants and florivores have been relatively little studied (Burgess, 1991). This is a promising arena for investigation because the flowers of many plants are fed upon by specialized or generalized florivores, and florivores and pollinators are likely to generate conflicting selective pressures on the flowers of many species. The nature and strength of these pressures are likely to vary with plant densities and geographical ranges.

Comparative information is an essential component of data collected to test predictions of traits expected to be favoured under conditions of low local population density. A successful approach requires precise predictions based on the traits of the focal species and data from enough closely-related species to yield statistically valid comparisons. Genera with large number of species or lineages with many genera are the best groups for comparative studies.

In addition to the results of broad, interspecific comparisons, data are needed from analyses of intraspecific variations. Because all species are rare somewhere, and local commonness and rarity may have persisted for evolutionarily significant time spans, intraspecific comparisons can demonstrate the influence of natural selection on plant populations living at different densities.

Rarity should influence the traits of animals, but the evolutionary consequences of rarity among motile animals are difficult to predict because individuals of rare species are able to aggregate for purposes of cooperative hunting, mating and defence. Ability to aggregate may be enhanced by long-distance signalling, but individuals of common species may often find themselves isolated from conspecifics, with the result that long-distance signalling may be advantageous for both rare and common species. Therefore, the demographic and population genetic consequences of recently imposed rarity may be of greater conservation significance among animals than they are among plants.

In contrast, plants and sessile animals are likely to exhibit many traits that have evolved in relation to population densities. Identifying those traits that evolve when population densities are low, and determining why they evolve, is of great theoretical interest because ecological knowledge is based primarily on the study of common species where they are common. In addition, if common species have traits that favour their success under conditions of high densities, but do not function well under conditions of low population densities, many plant species may be particularly vulnerable when their populations are greatly reduced by human activities. Deliberate intervention may be necessary to prevent those species from becoming extinct. Evolutionary insights are needed to guide appropriate intervention.

ACKNOWLEDGEMENTS

Clement Hamilton generously shared with me his extensive knowledge of *Psychotria*. My ideas about rarity have been developed over the years as a result of an ongoing dialogue with William Kunin, and I am often unable to remember which ideas are his and which are mine. Helpful comments on a draft of the manuscript were provided by William Kunin, Michael Rosenzweig and Douglas Schemske. My research on BCI was supported by a Regents' Fellowship from the Smithsonian Tropical Research Institute and sabbatical leave salary from the University of Washington.

REFERENCES

Anderson, W.R. (1973) A morphological hypothesis for the origin of heterostyly in the Rubiaceae. *Taxon*, **42**, 537–542.

Andrewartha, H.G. and Birch, L.C. (1954) *The Distribution and Abundance of Animals*. University of Chicago Press, Chicago.

Ashman, T.-L. and Schoen, D.J. (1994) How long should flowers live? *Nature* (London), **371**, 788–791.

Baker, H.G. (1958) Studies in the reproductive biology of West African Rubiaceae. *Journal of the West African Science Association*, **4**, 9–24.

Bannister, M.H. (1965) Variation in the breeding system of *Pinus radiata*, in *The Genetics of Colonizing Species*, (eds H.G. Baker and G.L. Stebbins), Academic Press, New York, pp. 353–374.

Barrett, S.C.H. (1989) Mating system evolution and speciation in heterostylous plants, in *Speciation and Its Consequences* (eds D. Otte and J.A. Endler), Sinauer Associates, Sunderland, MA, pp. 257–283.

Barrett, S.C.H. and Richards, J.H. (1990) Heterostyly in tropical plants. *Memoirs of the New York Botanical Garden*, **55**, 35–61.

Bawa, K.S. and Beach, J.S. (1983) Self-incompatibility systems in the Rubiaceae of a tropical lowland wet forest. *American Journal of Botany*, **70**, 1281–1288.

Bazzaz, F.A., Chiariello, N.R., Coley, P.D. and Pitelka, L.F. (1987) Allocating resources to reproduction and defense. *BioScience*, **37**, 58–67.

Beach, J.H. and Bawa, K.S. (1980) Role of pollinators in the evolution of dioecy from distyly. *Evolution*, **34**, 1138–1142.

Berenbaum, M. (1981a) Effects of linear furanocoumarins on an adapted specialist insect (*Papilio polyxenes*). *Ecological Entomology*, **6**, 345–351.

Berenbaum, M. (1981b) Patterns of furanocoumarin distribution and insect herbivory in the Umbelliferae: plant chemistry and community structure. *Ecology*, **62**, 1254–1266.

Brown, J.H. (1984) On the relationship between abundance and distribution of species. *American Naturalist*, **124**, 255–279.

Burgess, K.H. (1991) Florivory: The ecology of flower feeding insects and their host plants. PhD Thesis, Harvard University.

Cartar, R. and Dill, L. (1990) Why are bumblebees risk-sensitive foragers? *Behavioral Ecology and Sociobiology*, **26**, 121–127.

Cates, R.G. and Orians, G.H. (1975) Successional status and the palatability of plants to generalist herbivores. *Ecology*, **56**, 410–418.

Chew, F.S. (1975) Coevolution of pierid butterflies and their cruciferous food plants. I. The relative quality of available resources. *Oecologia*, **20**, 117–127.

Coley, P.D., Bryant, J.P. and Chapin, F.S. III (1985) Resource availability and plant antiherbivore defense. *Science*, **230**, 895–899.

Croat, T.B. (1978) *Flora of Barro Colorado Island*, Stanford University Press, Stanford.

Darwin, C. (1859) *On the Origin of Species*, Murray, London.

Dajoz, I., Till-Bottraud, I. and Gouyon, P.-H. (1991) Evolution of pollen morphology. *Science*, **253**, 66–68.

Dodson, C.H. (1962) The importance of pollination in the evolution of the orchids of tropical America. *American Orchid Society Bulletin*, **31**, 525–534, 641–649, 731–735.

Donoghue, M. J. and Doyle, J. A. (1989) Phylogenetic analysis of angiosperms and the relationships of Hamamelidae, in *Evolution, Systematics, and Fossil History of the Hamamelidae*, Vol. 1 (eds P.R. Crane and S. Blackmore), Clarendon Press, Oxford, pp. 17–45.

Erdtman, G. (1966) *Pollen Morphology and Plant Taxonomy*, Hafner, New York.

Feeny, P. (1976) Plant apparency and chemical defense. *Recent Advances in Phytochemistry*, **10**, 1–40.

Feinsinger, P., Murray, K.G., Kinsman, S. and Busby, W.H. (1986) Floral neighborhood and pollination success in four hummingbird-pollinated cloud forest plant species. *Ecology*, **67**, 449–464.

Fiedler, P.L. (1987) Life history and population dynamics of rare and common mariposa lilies (*Calochortus* Pursh: Liliaceae). *Journal of Ecology*, **75**, 977–995.

Futuyma, D.J. (1976) Food plant specialization and environmental predictability in Lepidoptera. *American Naturalist*, **110**, 285–292.

Gass, C.L. and Roberts, W.M. (1992) The problem of temporal scale in optimization: three contrasting views of hummingbird visits to flowers. *American Naturalist*, **140**, 829–853.

Hamilton, C.W. (1989a) A revision of Mesoamerican *Psychotria* subgenus *Psychotria* (Rubiaceae), part 1: Introduction and species 1–16. *Annals of the Missouri Botanical Garden*, **76**, 67–111.

Hamilton, C.W. (1989b) A revision of Mesoamerican *Psychotria* subgenus *Psychotria* (Rubiaceae), part 2: Species 17–47. *Annals of the Missouri Botanical Garden*, **76**, 386–429.

Hamilton, C.W. (1989c) A revision of Mesoamerican *Psychotria* subgenus *Psychotria* (Rubiaceae), part 3: Species 48–61 and appendices. *Annals of the Missouri Botanical Garden*, **76**, 886–916.

Hamilton, C.W. (1990) Variations on a distylous theme in Mesoamerican *Psychotria* subgenus *Psychotria* (Rubiaceae). *Memoirs of the New York Botanical Garden*, **55**, 62–75.

Hanski, I. (1982) Dynamics of regional distribution: the core and satellite species hypothesis. *Oikos*, **38**, 210–221.

Hanski, I., Kouki, J. and Halkka, A. (1993) Three explanations of the positive relationship between distribution and abundance of species, in *Species Diversity in Ecological Communities* (eds R.E. Ricklefs and D. Schluter), University of Chicago Press, Chicago, pp. 108–116.

Harper, K.T. (1979) Some reproductive and life history characteristics of rare plants and implications for management. *Great Basin Naturalist Memoirs*, **3**, 129–137.

Hoekstra, F.A. (1983) Physiological evolution in angiosperm pollen: possible role of pollen vigor, in *Pollen: Biology and Implications for Plant Breeding* (eds D.L. Mulcahy and E. Ottaviano), Elsevier Biomedical, New York, NY, pp. 35–41.

Hubbell, S.P. (1979) Tree dispersion, abundance, and diversity in a tropical dry forest. *Science*, **203**, 1299–1309.

Jennersten, O. (1988) Pollination in *Dianthus deltoides* (Caryophyllaceae): effects of habitat fragmentation on visitation and seed-set. *Conservation Biology*, **2**, 359–366.

Karron, J.D. (1987) A comparison of levels of genetic polymorphism and self-compatibility in geographically restricted and widespread plant congeners. *Evolutionary Ecology*, **1**, 47–58.

Kunin, W.E. (1991) Few and far between: plant population density and its effects on insect–plant interactions. PhD Thesis, University of Washington, Seattle.

Landa, K. and Rabinowitz, D. (1983) Relative preferences of *Arphia sulphurea* (Orthoptera: Acrididae) for sparse and common prairie grasses. *Ecology*, **64**, 392–395.

Levin, D.A. (1995) Plant outliers: an ecological perspective. *American Naturalist*, **145**, 109–118.

Macior, L.W. (1978.) The pollination ecology and endemic adaptation of *Pedicularis furbishiae* S. Wats. *Bulletin of the Torrey Botanical Club*, **105**, 268–277.

Martin, F.W. (1967) Distyly, self-incompatibility, and evolution in *Melochia*. *Evolution*, **21**, 493–499.

McNeill, J. and Crompton, C.W. (1978) Pollen dimorphism in *Silene alba* (Caryophyllaceae). *Canadian Journal of Botany*, **56**, 1280–1286.

Mehrhoff, L.A. III (1983) Pollination in the genus *Isotria* (Orchidaceae). *American Journal of Botany*, **70**, 1444–1453.

Mulcahy, D.L. (1974) Correlation between speed of pollen tube growth and seedling height in *Zea mays* L. *Nature* (London), **249**, 491–493.

Mulcahy, D.L. and Mulcahy, G.B. (1975) The influence of gametophytic competition on sporophytic quality in *Dianthus chinensis*. *Theoretical and Applied Genetics*, **46**, 277–280.

Murawski, D.A., Hamrick, J.L., Hubbell, S.P. and Foster, R.B. (1990) Mating systems of two bombacaceous trees of a Neotropical forest. *Oecologia*, **82**, 501–506.

Nilsson, L.A. (1992) Orchid pollination biology. *Trends in Ecology and Evolution*, **7**, 255–259.

Noss, R.E. and Cooperrider, A.Y. (1994) *Saving Nature's Legacy*, Island Press, Covelo, CA.

Oostermeijer, J.G., Den Nijs, J.C.M., Raijmann, L.W.E.L. and Menken, S.B.J. (1992) Population biology and management of the marsh gentian (*Gentiana pneumonanthe* L.), a rare species in the Netherlands. *Botanical Journal of the Linnean Society*, **108**, 117–130.

Ornduff, R. (1966) The origins of dioecism from heterostyly in *Nymphoides* (Menyanthaceae). *Evolution*, **20**, 309–314.

Preston, F.W. (1948) The commonness, and rarity, of species. *Ecology*, **29**, 254–283.

Primack, R.B. (1985) Longevity of individual flowers. *Annual Review of Ecology and Systematics*, **16**, 15–37.

Rabinowitz, D. (1981) Seven forms of rarity, in *The Biological Aspects of Rare Plant Conservation* (ed. H. Synge), John Wiley and Sons, New York, pp. 205–217.

Rabinowitz, D. and Rapp, J.K. (1981) Dispersal abilities of seven sparse and common grasses from a Missouri prairie. *American Journal of Botany*, **65**, 616–624.

Rabinowitz, D, Rapp, J.F., and Dixon, P.M. (1984) Competitive abilities of sparse grass species: means of persistence as a cause of abundance. *American Journal of Botany*, **68**, 1144–1154.

Rabinowitz, D., Rapp, J.K., Cairns, S. and Mayer, M. (1989) The persistence of rare prairie grasses in Missouri: environmental variation buffered by reproductive output of sparse species. *American Naturalist*, **134**, 525–544.

Real, L.A. (1981) Uncertainty and pollinator–plant interactions: the foraging behavior of bees and wasps on artificial flowers. *Ecology*, **62**, 20–26.

Rhoades, D.F. and Cates, R.G. (1976) Toward a general theory of plant antiherbivore chemistry. *Recent Advances in Phytochemistry*, **10**, 168–213.

Richards, A.J. (1986) *Plant Breeding Systems*, Allen and Unwin, London.

Ricklefs, R.E. and Cox, G.W. (1972) Taxon cycles in the West Indian avifauna. *American Naturalist*, **106**, 195–219.

Sih, A. and Baltus, M.-S. (1987) Patch size, pollinator behavior, and pollination limitation in catnip. *Ecology*, **68**, 1679–1690.

Smyth, C.A. and Hamrick, J.L. (1984) Variation in estimates of outcrossing in musk thistle populations. *Journal of Heredity*, **75**, 303–307.

Snow, A.A. and Spira, T.P. (1991) Pollen vigor and the potential for sexual selection in plants. *Nature* (London), **352**, 766–797.

Squillace, A.E. and Krause, J.R. (1963) The degree of natural selfing in slash pine as estimated from albino frequencies. *Silvae Genetica*, **12**, 46–50.

Terborgh, J. and Winter, B. (1980) Some causes of extinction, in *Conservation Biology: an evolutionary–ecological perspective* (eds M.E. Soulé and B.A. Wilcox), Sinauer Associates, Sunderland, MA, pp. 119–133.

Thornhill, R. and Alcock, J. (1983) *The Evolution of Insect Mating Systems*, Harvard University Press, Cambridge, MA.

Wheelwright, N.T. and Orians, G.H. (1982) Seed dispersal by animals: contrasts with pollen dispersal, problems of terminology, and constraints on coevolution. *American Naturalist*, **119**, 402–413.

Wilson, E.O. (1961) The nature of the taxon cycle in the Melanesian ant fauna. *American Naturalist*, **95**, 1659–193.

Wolf, L.L. and Hainsworth, F.R. (1983) Economics of foraging strategies in sunbirds and hummingbirds, in *Behavioral Energetics: The Cost of Survival in Vertebrates* (eds E.P. Aspey and S.I. Lustic), Ohio State University Press, Columbus, OH, pp. 223–264.

12 Rarity and evolution: some theoretical considerations

Robert D. Holt

12.1 INTRODUCTION

Does rarity matter in evolution? This question has no simple answer, because the terms 'rarity' and 'evolution' each denote complex rather than simple things. Evolution, in its broadest sense, incorporates all aspects of the origin and maintenance of biotic diversity by processes within and among species (e.g. selection and speciation). Rarity is likewise a compound concept, usually involving comparisons among populations or species (Gaston, 1994).

Gaston (1994) has proposed that the term 'rarity' be used in a relative sense to denote (say) the least abundant 25% of species in a community. This proposition has its merits, but other usages are also reasonable. For instance, rarity might indicate the subset of a rank-ordered list of species, starting with the least common, which collectively comprise 25% of total community abundance (this subset often includes a majority of species). Or, a species may be deemed 'rare' assessed against an absolute criterion, such as one of those that emerge from models of demographic stochasticity or Allee effects, or by its likelihood of being observed in a defined sample. For instance, a temperate-zone bird-watcher after a day of meagre birding in the Atacama Desert could reasonably conclude that all bird species there are 'rare'; with Gaston's definition, there are by convention exactly as many rare species there as anywhere else.

In this chapter, when considering a local community, the word 'rarity' will be shorthand for 'low population size'. When considering large biogeographical areas, 'rarity' may denote either a small total abundance for a species or a small range size (e.g. number of sites occupied), depending on context; 'spatial rarity' will be used to indicate the latter aspect of rarity.

The Biology of Rarity. Edited by William E. Kunin and Kevin J. Gaston. Published in 1997 by Chapman & Hall, London. ISBN 0 412 63380 9.

'Rarity' may be related to 'evolution' in many ways; a given component of rarity may be irrelevant to some aspects of evolution, yet critical for others (Ridley, 1993). The aim in this chapter is to provide an overview of our current understanding of how two aspects of 'rarity' – small local population size and small range size – influence evolutionary dynamics. The emphasis is on evolutionary theory, rather than empirical patterns (thus complementing other chapters in this volume, e.g. Chapters 10, 11) and particularly on unresolved or poorly explored conceptual issues. Although the consequences of rarity for microevolution are examined most closely here, some implications of rarity for macroevolution are also touched on.

To place current issues in a historical context, a capsule history of ideas about the role of population size in evolution is first presented. Next, we take a hierarchical approach to analysing the evolutionary implications of rarity. Some evolutionary implications of population size in a closed local community (i.e. where immigration is absent) are considered. After some comments on differences between ecological and genetical measures of population size, the influence of population size on neutral evolution, adaptive evolution, and sympatric speciation is discussed. The chapter then looks at the relation of population size to adaptive evolution in local communities open to immigration. Finally, it examines the more complex (but realistic) issue of species rarity at broader spatial scales, encompassing many local communities, where one must consider the effects of both total population size and range size on evolution.

One message of this chapter is that rarity may have differing evolutionary roles at different spatial, temporal and comparative scales of analysis. Many different kinds of comparisons are possible. One could compare evolution of two species (one common, the other rare) at the same site; or of pairs of populations at different sites within the same species' range; or of species as a whole over their entire ranges; or of all rare species vs. all common species (or all rare populations of a species vs. all common populations). A second message that runs through the chapter is the importance of grounding evolutionary questions in a firm understanding of population dynamics.

12.1.1 Historical context

Ideas about the role of rarity in evolution have long been central to conceptual controversies in evolutionary biology. Charles Darwin (1859 [1964: 53; see also 105, 107]) argued that large and widely distributed populations are the locus of most significant evolution:

> It is the most flourishing . . . , the most dominant species – those which range widely over the world, are the most diffused in their own country, and are the most numerous in individuals, – which oftenest produce well-marked varieties, or . . . incipient species.

As summarized crisply by Leigh (1986: 192–193), Ronald Fisher likewise believed that:

> large populations allow more mutation, discriminate selective advantages more precisely, and retain a larger proportion of favourable mutations, than do smaller ones. In practice, large populations vary most. Thus, one expects a smaller number of large populations to be increasing at the expense of a larger number of small ones. Accordingly, the study of evolution should focus on abundant species, which are the fount of adaptation.

This is Darwin's worldview, dressed up in modern genetics.

Many evolutionists have argued against the evolutionary centrality of large populations. Wright's shifting balance theory (Wright, 1931, 1977) highlighted a potential creative role for drift, permitting small populations to move between adaptive peaks. The disagreement between Fisher and Wright on the role of population size is one of the oldest controversies in population genetics (Maynard Smith, 1983a) and arises from sharply different views of epistasis (Wade, 1992).

The shifting balance theory (Eldredge, 1985) led many evolutionists to claim that major adaptive evolution is less likely in large populations. Mayr (1982: 604), for instance, asserts that 'large, widespread populations – in fact all more populous species – are evolutionarily inert'. The assumption that rarity promotes evolution has figured prominently in the debate over 'stasis' in the fossil record (e.g. Eldredge and Gould, 1988; Stanley, 1990).

Other evolutionists (e.g. Maynard Smith, 1983b) doubt that large population size hampers evolution: 'The supposed evolutionary inertness of large populations . . . is based on no evidence' (Coyne, 1994). Stasis, for instance, may reflect stabilizing selection (Charlesworth et al., 1982). Theoretical studies of the interplay of drift, selection and gene flow qualitatively agree with Darwin and Fisher: adaptive evolution should often be more rapid and precise in larger populations (Holt, 1987).

Analysis of the shifting balance theory is a highly active area of theory (e.g. Barton and Rouhani, 1992). The jury is still out on the importance of this process (Crow et al., 1990; Wade, 1992). 'Punctuational' changes need not imply an important role for rarity: peak shifts may occur in large populations, given transient environmental changes or increased genetic variance (Kirkpatrick, 1982; Merrell, 1994).

Ernst Mayr's theory of peripheral speciation emphasizes episodes of very low abundance in isolated populations (e.g. 'founder effects': Provine, 1989; Barton, 1989). By contrast, the standard Darwinian view is that speciation emerges as a byproduct of adaptive evolution proceeding independently in isolated populations. Maynard Smith (1989: 280) observes that even if small peripheral populations do exhibit more rapid speciation, this may not be because of their population size. If, within a species, populations at low density more often experience strong directional selection than do more abundant populations and peripheral populations tend to be isolated from gene flow, small peripheral populations may incidentally experience factors that jointly facilitate speciation, irrespective of their rarity. Local rarity may thus be correlated with speciation, but not causally relevant.

To conclude this historical sketch, it is clear that evolutionists differ sharply in their views on the role of rarity (in its meaning of 'low population size') in evolution. The lack of unanimity on such a fundamental aspect of the history of life is rather astonishing.

12.2 EVOLUTION IN LOCALLY RARE POPULATIONS

12.2.1 Closed communities

To clarify the interplay of rarity and evolution, it is helpful to begin with the simplest case, namely a closed community which after its initial establishment is closed to further immigration or colonization. In a local community, a species is 'rare' if it has low population size, relative to other species in the community. Rarity may influence many aspects of evolution in a closed community, from 'arms races' in specialist predator–prey interactions (Schaffer and Rosenzweig, 1978) to long-term responses to environmental change (Holt, 1990).

(a) A methodological difficulty

A significant gap in relating evolutionary theory to field data is that the relevant population size is not census abundance, N_{cens}, but 'effective population size', N_e, which governs both drift and the amount of variation available for selection (Hedrick, 1983; Crawford, 1984; Holt, 1987). Even in apparently thriving populations, N_e may often be low enough for significant drift (for examples, see Lande, 1979; Husband and Barrett, 1992).

Temporal fluctuations in abundance imply $N_e < N_{cens}$ (Hedrick, 1983; Begon, 1992). N_e is approximately the time-averaged harmonic mean of population size (Crawford, 1984), a quantity heavily biased toward low N. For example, a population that spends a fraction q of generations at N_{low}, and the remainder at huge numbers, has $N_e < N_{low}/q$ (Holt, 1987). A typically abundant species with occasional excursions to low densities may have a lower N_e than a rare species with stable dynamics. Crawford (1984) reports one to two order-of-magnitude differences between census and effective population sizes in arthropods, but smaller (two-fold) differences in large vertebrates (possibly reflecting differences in the scale of population fluctuations in these two groups).

Variation maintained in a mutation-drift balance increases with N_e (Hedrick, 1983), as does genetic variance of neutral quantitative characters

(Lynch, 1994). Common species indeed often harbour more neutral genetic variation than rare species (Chapter 10; Nei and Grauer, 1984). For instance, Hamrick *et al.* (1991) compared allozyme diversities of common and uncommon tree species on Barro Colorado Island, Panama, and found locally rarer species had lower genetic diversity.

The relevant time-series that determines N_e may be very long. Gillespie (1991: 54) notes that though species with population sizes in the range $10–10^3$ exhibit unusually low levels of heterozygosity and a positive correlation between population size and heterozygosity, for species exceeding 10^4 in population size heterozygosity is essentially independent of N. This could be a genetic signature of population fluctuations over long time-scales. Nei and Grauer (1984; see also Barrett and Kohn, 1991) argue that contemporaneous interspecific patterns of protein heterozygosity reflect Pleistocene bottlenecks. (By contrast, sensitivity to infrequent bottlenecks may not characterize quantitative traits, which have relatively high mutation rates: Lande, 1980.)

Pimm (1991) has observed that variance in abundance seems to increase, the longer one observes a population. Short-term studies are likely to overestimate N_e. There seem to have been no attempts to compare relative rankings of species in a community by N_{cens} and by N_e. Were the two rankings similar, this would suggest consistency in current rank-order abundances over evolutionarily significant time-scales.

Given that a local population persists, its abundance must be bounded away from zero. Populations with low average abundances should tend to have low temporal variance in abundance; populations with high average abundances can persist in the face of higher variance. This dynamical constraint suggests that populations may differ less in N_e than in average N_{cens}; this is because N_e is determined principally by population lows (and persisting populations cannot get too low in abundance and still persist), whereas average abundance is strongly influenced by population highs (which can vary widely without affecting the likelihood of persistence). It would be a useful theoretical exercise to assess systematically the effect of different patterns of population dynamics on N_e. In any case, gauging the relevance of local rarity to evolution requires consideration of the dynamical behaviour of populations – not just mean abundances.

(b) Rarity and rates of neutral evolution

Neutral evolution may matter on its own (e.g. in molecular evolution: Nei, 1987) or as a null model for phenotypic evolution (Lynch, 1994). A simple, striking prediction of the theory of neutral molecular evolution is that the rate of allelic substitution at a single locus is the per capita mutation rate, independent of N_e (Kimura, 1983). Unless the mutation rate varies with N_e, there should be no correlation across species between population size and the average rate of neutral evolution. Models of neutral phenotypic evolution likewise predict that expected between-population variance in mean genotypic values increases linearly with time (Lynch, 1994), independent of N_e.

Population size does not influence the rate of neutral evolution within a lineage. Does this imply rarity is irrelevant to neutral evolution in a local community? Not at all! In comparing a rare species with a common one, on average they should not differ in evolutionary rates. However, if at the level of the entire community we compare evolution in rare species (as a class) with common species (as a class) we must account for differences in species richness. In communities, different abundance classes typically differ in species richness.

Interspecific variation in abundance typically fits lognormal or logseries distributions (May, 1975; Tokeshi, 1993), which are right-skewed, with more moderately rare species than common species. Because mutation and drift are stochastic, species will vary in their realized rates of evolution. If an evolutionary 'novelty' is defined as a phenotypic change greater than a specified amount (compared with a species' ancestral condition), one is more likely to observe a novelty first arising in moderately rare species than in common species, simply because there are more of the former, providing a larger sample for the stochastic evolutionary process. Arnold and Fristrup (1982) make a similar point about the evolutionary scope of clades differing in species richness.

So, given neutral evolution, there may be emergent macroevolutionary effects of population size at the community level, even if population size has no average effect on the rate of microevolution in any species. The utility of this insight rests on the largely untested potential for neutral models to explain interesting patterns of phenotypic evolution. In most cases, phenotypic evolution is obviously driven by selection.

(c) Local rarity and adaptive evolution: direct effects

Rarity can directly affect adaptive evolution if the rank-order fitness of phenotypes or genotypes varies with population size. From the perspective of an individual organism, population abundance is one of many environmental factors that may influence its fitness (Chapter 9). There are two basic ways this can happen, encapsulated in the distinction between 'hard' and 'soft' selection (Hedrick, 1983). Standard models of density-dependent selection (e.g. Begon, 1992; Charlesworth, 1994) analyse selection at those life stages which determine abundance. The outcome of selection in these models is to increase carrying capacity ('hard' selection). Rare populations which experience hard selection and persist should, as a byproduct, become less rare (Gomulkiewicz and Holt, 1995).

But rank-order fitness can vary with population size, without reciprocal effects of selection on population size ('soft' selection). Orians (Chapter 11) argues that the reproductive syndromes of rare plants may differ systematically from common plants (e.g. in being less dependent on specialist pollinators; see also Kunin and Gaston, 1993). Such adaptive consequences of rarity need not make populations more abundant as a byproduct of selection.

The literature of evolutionary ecology is replete with verbal arguments and formal theoretical models exploring coevolutionary consequences of population size (e.g. Rosenzweig *et al.*, 1987; Begon, 1992). Two examples suffice to illustrate these ideas:

- Populations rare because of interspecific competition may evolve enhanced interspecific competitive ability, whereas competitive dominants may be selected for improved intraspecific competitive ability (Pimentel *et al.*, 1965; Aarssen, 1983).
- If predators consume substitutable prey, selection on attack rates should be biased toward abundant prey types (Holt, 1977).

Such effects are doubtless important. However, within a community, rare species are a highly heterogeneous lot. Local rarity has myriad causes (Gaston, 1994); many distinct evolutionary syndromes can thus be associated with rarity. It is unlikely anything very sensible can be said about the adaptive importance of rarity, *per se*, outside carefully delimited comparative contexts (e.g. between sympatric congeners, as in Orians' example in Chapter 11).

Such comparisons often benefit from being cast within an explicit or implicit model for population dynamics. Let us assume each species in a local community follows a generalized logistic growth model:

$$\frac{dN_i}{dt} = [r_i - d_i N_i - m_i(t)]N_i$$

where N_i is the local abundance of species i, r_i its intrinsic rate of increase, d_i is the strength of density-dependence species i exerts on itself, and $m_i(t)$ is temporally varying density-independent mortality imposed on species i (e.g. from episodic disturbances). Species i deterministically persists if:

$$r_i - <m>_i > 0$$

where $<m>_i$ is the time-averaged mortality experienced by species i. The time-averaged abundance of population size (Levins, 1979) is:

$$N^*_i = (r_i - <m>_i)/d_i.$$

A species may be rare in a closed community because of low r_i (e.g. the habitat is near the edge of its fundamental niche), or because density dependence is strong (high d_i), or because of frequent, severe disturbances (high $<m>_i$). These very different dynamical reasons for rarity could imply quite distinct selective regimes.

Yet the relation between the ecological environment which determines population dynamics and the selective environment which generates evolutionary dynamics is complex and can at times be counter-intuitive. Ecologists often casually assume that as a given factor becomes more important in limiting or regulating population size, that factor automatically becomes more important in selection. The folk wisdom is that 'evolution works hardest where the shoe pinches worst' (W. Kunin, personal communication). For a spatially homogeneous system, the consumer-resource model (MacArthur, 1972) can be used to illustrate two distinct classes of counter-example to this intuition.

Let C_i be the density of consumer species i, and R_i the availability of its required resource. We assume each consumer species exploits its own exclusive resource (i.e. no interspecific competition among consumers) and grows according to:

$$dC_i/dt = C_i (a_i b_i R_i - m_i)$$

where a_i is the per unit rate of uptake, b_i converts consumption into consumer births and m_i measures density-independent mortality. Assume resource dynamics are given by:

$$dR_i/dt = I_i - u_i R_i - a_i R_i C_i$$

where I_i measures resource input, and u_i the rate of loss other than to consumption. At equilibrium, resource and consumer abundances are, respectively:

$$R_i^* = m_i/a_i b_i$$

and:

$$C_i^* = b_i I_i/m_i - u_i/a_i$$

In a local community, as noted above, one consumer population may be rarer than another for any of a number of distinct dynamical reasons. In this resource–consumer system, a consumer may be rare because it experiences a high mortality rate (high m_i), or is ineffective at acquiring the resource (low a_i), or has a low-quality resource (low b_i). Alternatively, a consumer population may be rare because its resource has a low input rate (low I_i) or a high loss rate (high u_i). The former class of reasons focuses on parameters which enter directly into the consumer's own growth equation. By contrast, the latter class of reasons for rarity involve parameters of the resource, not parameters of the consumer population. Only the former parameters are directly accessible to selection in the consumer.

Now assume that consumers vary intraspecifically in exploitative ability, and that such variation affects only the attack rate a_i. For a moment, assume that the consumer does not influence resource availability. The strength of selection for an increase in attack rate is given by:

$$d/da_i [(1/C_i)dC_i/dt] = b_i R_i.$$

The greater the R_i, the greater is the strength of selection for increasing the rate of exploitation on the resource. Conversely, decreased resource availability reduces the strength of selection for improved exploitative abilities (Abrams, 1986, elaborates this theme). As food becomes more important in limiting population growth (i.e. lower R_i), the strength of selection for improved food acquisition can become weaker, not stronger.

If we assume that there is a trade-off between the ability to acquire the resource and ability to withstand mortality factors, this may imply the existence of an optimal attack rate. Formally, let:

$$M_i = M + m'_i(a_i)$$

where M is fixed mortality and $m'_i(a_i)$ is the additional mortality that accrues because of resource consumption. Assume $dm'_i/da_i > 0$ and $d^2m'_i/da_i^2 > 0$ (e.g. more intense foraging has risks which increase at a faster than linear rate). The optimal phenotype is found from:

$$d/da_i \, [(I/C_i)dC_i/dt] = 0$$

or:

$$b_iR_i = dm'_i/da_i.$$

The lower the R_i, the lower is the optimal uptake rate (and the death rate). In a resource-poor environment, consumers may evolve toward lower exploitation rates.

Now assume that the resource–consumer interaction is in demographic equilibrium, so consumption limits resource availability. After setting $R_i = R_i^*$, the optimal uptake rate is found from:

$$m_i/a_i = dm'_i/da_i,$$

an expression which is independent of the parameters I_i, u_i and b_i. Differences between species in these parameters affect local consumer abundance but do not influence the optimal consumer phenotype. Moreover, an increase in fixed mortality (M) increases R_i^*, which indirectly increases the optimal rate of resource exploitation, even at the expense of greater net mortality (m_i).

The above model illustrates several general points. First, population size can be influenced by many factors not part of the selective environment directly experienced by individuals. In a local community, one species may be rare and another common (e.g. because of differences in the renewal rates of required resources), without any necessary qualitative difference in the selective environments experienced by individuals.

Second, increasing the magnitude of a limiting factor need not automatically evoke a selective response that tends to counter that environmental change (given the existence of genetic variation). Elsewhere (Holt, 1996a,b, and see below) it has been similarly shown that in populations with a source-sink structure, as a population becomes increasingly restricted to the source the strength of selection favouring adaptive evolution in the sink becomes weaker, not stronger. A fundamental problem in understanding the evolutionary implications of local rarity is thus articulating the relation between factors influencing local population size, and those influencing adaptive evolution. The above model and the results reported in Holt (1996b) illustrate that this relation may be subtle, non-obvious and often counter-intuitive.

Third, the particular conclusions reached above clearly apply to a particular model. Change the assumptions of the model, and the resulting predictions are likely to change. The aim in this chapter is not to champion the specific conclusions of this model, but rather to suggest that discussions of the relation of rarity to evolution need to be cast in the context of explicit population dynamic models, and that intuitive notions about the mapping of limiting factors on to selective factors may at times be off-base.

(d) Local rarity and adaptive evolution: indirect effects

Population size indirectly influences adaptive evolution via control of the supply of heritable variation, and by altering the relative importance of selection and drift. Theory predicts that large populations generate more novel mutations per generation and retain more variation against the influence of drift (Holt, 1987). Sometimes this is unimportant. For instance, in constant environments, if relative fitnesses can be described by a function with a single optimum, even weak selection can maintain populations near their optimal phenotype, except in quite small populations (Lande, 1980).

Wright's shifting balance theory suggests that small (but not too small) populations may have the greatest capacity for adaptive evolution. However, this theory works best when considering conspecific populations with the same adaptive surface. In the absence of a universal metric for describing adaptation across species in a community (which seems improbable *a priori*) it may be difficult to attach much meaning to the claim that rare species are more, or less, precisely adapted than are common species.

The metaphor of adaptive landscapes may not be very useful for natural enemy–victim interactions, which typically involve frequency-dependent selection. Seger (1992) argues that natural enemy–victim systems are prone to evolutionary instability. Cyclic or chaotic dynamics in gene frequencies can force some allelic frequencies to drop to low values, fostering the loss of variation in small populations. Long-term evolutionary instability in enemy–victim systems (if it occurs at all: Rosenzweig *et al.*, 1987) requires both species to maintain variation, so should be most likely if both species sustain large effective population sizes.

One indirect evolutionary effect of rarity which has been recently recognized is that small populations tend to harbour larger loads of deleterious mutations than do large populations, and so may be more prone to extinction via 'mutational meltdown' (Lynch and Gabriel, 1990; Lande, 1994). This maladaptive evolutionary process can lead to the eventual disappearance from a local community of species which were initially adapted to local conditions, leading to differential extinctions of rare species far beyond the expectations of demographic stochasticity alone (Lande, 1994).

(e) Local rarity in temporally varying environments

Environmental change can induce phases of directional selection as populations move towards new phenotypic optima (Grant and Grant, 1989). Directional selection experiments show that both initial rates of response to selection and ultimate selection limits increase with population size, qualitatively matching theory (Hill and Caballero, 1992). As noted above, temporal variation in abundance depresses N_e. This enhances drift, on the one hand making shifting balance evolution more likely, but on the other depleting the pool of variation. Population fluctuations may also lead to temporal variation in selection, which in turn can make peak shifts more likely without drift (Kirkpatrick, 1982; Merrell, 1994). The overall effect of transient phases of local rarity on adaptive evolution, taking into account all these factors, is not at all clear.

Recently, theoreticians have explored evolutionary dynamics of populations in changing environments (e.g. Holt, 1990; Lynch and Lande, 1993; Gomulkiewicz and Holt, 1995). In a closed community, evolution is the only feasible route to survival given sufficiently great environmental change. Rare species may be especially vulnerable to extinction in changing environments. Low N_e implies low variation, reducing the rate of response to selection in a novel environment. There is a roughly log-linear increase in the permissible rate of environmental change – beyond which extinction is inevitable – with increasing N_e (Lynch and Lande, 1993).

After severe environmental change, species should decline in number even as they adapt. Given a rare and a common species with the same genetic variation, the rare species is more likely to dip to densities where extinction is likely due to demographic stochasticity (Gomulkiewicz and Holt, 1995). Though large populations are no buffer against large environmental change, even small changes can potentially endanger rare populations – including those with the genetic wherewithal to adapt to changed circumstances.

These conclusions concern pairwise comparisons of common and rare species. Interspecific phenotypic variation usually greatly exceeds intraspecific variation. In a closed local community, the only species likely to survive a radical environmental change may be rare species – simply because a greater initial range of phenotypes was available among rare species, increasing the likelihood that at least one species was preadapted to the change. Environmental change often leads to dramatic shifts in dominance, with initially rare species becoming abundant (Holt, 1995). Long-term community evolution may be dominated by the descendents of species which were rare at the time of large environmental change, merely because this class of species is relatively speciose.

(f) Rarity and sympatric speciation

Macroevolution in a closed community is likely to be largely driven by differential species extinction, which should be biased toward rarer community members. The only other macroevolutionary process which can operate in a closed community is sympatric speciation.

The most widely accepted mechanism for sympatric speciation is polyploid speciation. Hybridization between sexual species produces allopolyploids, a process that occurs reasonably often in plants (Grant, 1981). Considering all potential parental pairs of species, the rate of hybridization should be low for pairs of rare species, versus pairs with at least one common species. Indeed, in many plant-pollinator systems, rare species may be particularly vulnerable to hybridization with commoner species (W. Kunin, personal communication). This can lead to extinction, but also provides a source for polyploid speciation. There seems to be no formal theory bearing on this potential effect of population size on rates of allopolyploidy.

Autopolyploids are created by meiotic irregularities creating diploid gametes, which then combine. The rate of production of meiotic anomalies is likely to be proportional to population size. All else being equal, common species will be the most likely progenitors of autopolyploid species (Rosenzweig, 1995).

Other mechanisms for sympatric speciation are more controversial and depend upon density-dependent competition and heterogeneous habitats or resources (e.g. Rosenzweig, 1995; Wilson, 1989). Population size is not well correlated with local niche breadth (a measure of heterogeneity in habitat or resource use: Gaston, 1994), so there is no obvious reason to expect local rarity to affect the likelihood of these modes of sympatric speciation.

12.2.2 Open communities

Now consider a local community open to immigration, but without reciprocal ecological or evolutionary effects on the source pool for immigrants. Communities on oceanic islands often match this scenario reasonably well. Spatial openness in a local community has both intraspecific and interspecific effects. First, immigration into established populations permits gene flow to interact with selection, mutation and drift. Second, local species richness can be enhanced. Species may occur which could not persist in a closed community, either because they are vagrants (Gaston, 1994) or sink populations (Pulliam, 1988; Holt, 1993); or because they can re-colonize following local extinction; or because spatial openness permits ways of life impossible in closed communities, increasing the 'effective niche space' of the community (Holt, 1993).

Suggestive evidence of the evolutionary importance of rarity in open communities comes from analyses of island biotas. Diamond (1984, Figure 17) reported that endemism of Pacific island birds (at the species level or higher) increases with both island area and distance from continental sources. The following paragraphs build on Diamond's explanations for this pattern to identify several distinct evolutionary influences of rarity, possibly revealed in these area and distance effects in endemism.

Assume that, in the source, species have fixed properties. All colonization is from source to island. For species i, let c_i (A,d) be its colonization rate on to islands of given area A and distance d from the source, and e_i (A,d) its local extinction rate. In general (MacArthur and Wilson, 1967) colonization increases with area and decreases with distance; extinction decreases with area (a proxy for population size) and increases with distance. The fraction of islands in a given area–distance class with species i is p_i – the 'occupancy' of islands by species i (Hanski, 1991). Occupancy changes by:

$$\frac{dp_i}{dt} = c_i (1 - p_i) - e_i p_i \tag{12.1}$$

At equilibrium:

$$p_i^* = c_i/(c_i + e_i).$$

(a) Area effects on endemism

Populations on different islands of the same size and area are likely to differ in age (i.e. time since colonization). With constant extinction rates, at equilibrium all those populations within a given island area–distance class should exhibit an exponential distribution of ages. Assume that an evolutionary 'clock' starts ticking at colonization. If after T time units, divergence is sufficient to be deemed systematically relevant, a fraction $exp[-e_i T]$ of island populations should have survived long enough to become systematically distinct.

Among populations on different islands stemming from a given ancestral source species, those on larger islands should be observed to be more differentiated, for two reasons related to population size. First, they tend to be older (because they are larger in local abundance and so have lower extinction rates), and thus will have had more opportunity to become evolutionarily distinct (Diamond, 1984). Second, complementing this ecological effect, for the genetical reasons sketched above, larger populations may also adapt more rapidly to fixed or changing environments.

Immigration may also contribute to area effects on endemism. The number of immigrants per species per unit time into an island should scale with a linear dimension of island size (MacArthur and Wilson, 1967), while the resident population size should scale with this linear dimension squared. The relative contribution of immigrants to an island gene pool should thus decline with increasing island area, for islands equidistant from the source.

This will not affect neutral evolution; a very small amount of gene flow prevents neutral genetic differentiation, almost independent of population size (Hedrick, 1983). By contrast, gene flow may be more likely to constrain **adaptive** evolution in locally rare than in locally common populations, relative to a source population (e.g. Antonovics, 1976; Endler, 1977; Holt, 1987). Consider a population with discrete generations and viability selection at a diploid locus. $N(t)$ is the local population size at the start of generation t, $N'(t)$ population size following selection, and $i(t)$ the number of immigrants. Assume that immigrants are fixed for allele 2, and allele 1 is locally favoured. Let w_{ij} be the fitness of genotype ij; and assume both w_{11} and $w_{12} > w_{22}$. Over T generations, allele 1 increases when rare if:

$$\prod_{t=0}^{t=T-1} L(t) > (w_{22}/w_{12})^T$$

where $L(t) = N'(t)/[N'(t) + i(t)]$ is the fraction of breeding adults recruited locally, rather than by immigration (Nagylaki, 1979; Holt, 1987).

With constant abundances and immigration rates, persistence of the locally favoured allele requires $w_{12} > w_{22}(1 + i/N')$. Alleles with small positive effects on fitness are more likely to be retained if population size is large, relative to immigration. Given islands of area A, immigration scales as $A^{1/2}$ and population size as A; the selective advantage needed for retention of a locally favoured allele thus scales with $(1 + A^{-1/2})$. This implies that for a fixed immigration rate, rare (small island) populations are more vulnerable to gene flow. Moreover, if local population size fluctuates, the swamping effects of gene flow are magnified (Holt, 1987). The effect of demographic stochasticity looms larger in small populations, leading to fluctuations that further aggravate the effects of gene flow.

(b) Distance effects on endemism

The distance effect on endemism could reflect both within- and between-species consequences of rarity. Immigration increases population longevity (Brown and Kodric-Brown, 1977; Holt, 1993, and above), possibly increasing the scope for local evolution. However, as Diamond (1984) notes, such immigration also hampers local differentiation due to gene flow. The observed distance effect on endemism may imply that the genetic effect of gene flow (hampering differentiation) outweighs its ecological effect (enhancing local population persistence). In the above model, $i(t)$ decreases with distance; distant islands should be less subject to gene flow, permitting greater differentiation.

This within-species effect of distance on differentiation overlays a complementary between-species effect. Species vary greatly in extinction rates (often paralleling differences in local N). The species turnover predicted by island biogeography on close inspection reveals the churning of rare species (Williamson, 1981; Schoener and Spiller, 1987). Gaston (1994) suggests that many rare species in open communities may be 'vagrants', unable to persist without immigration and presumably prone to high extinction rates.

It is easy to show, using an elaboration of (12.1), that given hetero-geneous extinction rates, as one decreases overall colonization rates, island communities become biased towards species with low extinction rates, with a corresponding opportunity for more evolution. Averaged across species in a community, the mean longevity of island populations should increase with island distance. This interspecific effect compounds the intraspecific effect of a weakening of gene flow with distance, leading to the observed pattern of greater endemism on more distant islands.

(c) Local rarity as an indicator of evolutionary 'traps' in open communities

As noted above, many rare species in open communities may be vagrants. Why don't vagrant populations maintained by immigration evolve by natural selection so as to become responsible populations, self-sustaining in the local setting?

A 'vagrant' population that persists by immigration is a sink population (Pulliam, 1988; Holt, 1993). Theoretical models (Holt and Gaines, 1992; Holt and Gomulkiewicz, in press) suggest that the lack of adaptive evolution in sink populations may not be because gene flow hampers local selection, but instead simply reflects the fact that the vagrants are maladapted to the local environment in the first place. This can be seen using the above criterion for selection in the face of gene flow. Assume that at the start of generation t there are $N(t)$ individuals, all homozygous for allele 2, with absolute fitness in the local environment w_{22}. The population declines geometrically; after reproduction there are $N'(t) = N(t)w_{22}$ individuals. An aliquot of I individuals homozygous for allele 2 immigrates; after censusing there are $N(t+1) = N'(t) + I$ individuals. The equilibrial population size is $N = I/(1 - w_{22})$.

Now, introduce a very small number of individuals with allele 1 in heterozygous form. After substitution for N in the above criterion for a locally favoured allele to increase when rare, we find that in a sink population maintained by immigration the locally favoured allele 1 increases when rare if and only if $w_{12} > 1$. Note that this criterion for selection to enhance local adaptation is independent of the rate of immigration. Let $dw = w_{12} - w_{22}$ denote the effect of allele 1 upon individual fitness in the heterozygote. For $w_{22} < 1$, if dw is sufficiently small, $w_{12} < 1$. Hence, alleles of small positive effect on fitness cannot be favoured in a sink habitat. The evolution of adaptation by natural selection in a sink may require the availability of mutations of large effect (Holt and Gomulkewiecz, in press). If all available mutations have small effects on fitness, a sink habitat could remain a sink over extended time-scales, with no local adaptation.

Rare populations maintained by immigration may be observed not to adapt to the local environment, not quite because gene flow hampers local adaptation, but because rarity itself is a signature of low immigrant fitness. In this model, the influence of population size on the magnitude of gene flow disappears as a causal explanation for the lack of local adaptation in a rare sink population, once one properly accounts for the demographic as well as genetic effects of immigration (Holt and Gomulkiewicz, in press).

Hence, in an open community, rare species may be both less differentiated (from source populations) and less well adapted to the local environment, for several distinct reasons:

● In a community at equilibrium between colonization and extinction, rare species are more prone to extinction, and so on average are younger (the expected correlation between rarity and age may break down in non-equilibrial communities).
● If rare species persist, they are more vulnerable to gene flow (for a given rate of immigration).

In the latter case, local rarity is directly implicated in the lack of differentiation or adaptation. However, some species (vagrant populations in sink habitats) may be rare because they are maladapted and sustained by immigration; in this case, rarity is merely correlated with a lack of differentiation or adaptation, and is not its cause.

12.3 EVOLUTIONARY EFFECTS OF SPATIAL RARITY

Having examined in detail the evolutionary implications of rarity in first closed, then open, local communities, the next logical step is to move up in spatial scale to consider evolutionary effects of rarity for species distributed among several to many local communities. Entirely new evolutionary processes arise at larger spatial scales (e.g., group selection, allopatric speciation). The burgeoning literature on the evolutionary effects of population structure and metapopulation dynamics (e.g. Olivieri *et al.*,

1990; McCauley, 1993; Hastings and Harrison, 1994; Whitlock, ms.) will not be synthesized here; instead, let us highlight some general issues worthy of further attention.

Among species with comparable dispersal abilities, widespread species by definition have more local populations than do spatially rare species. For simplicity, consider species occupying discrete habitat patches, each patch potentially harbouring a population. We can describe a species' range with a population–abundance distribution, $p(n)$ (where local population size $= n$), which is a function giving the number of local populations falling into each abundance class. The total number of patches occupied is $p_T = \Sigma\ p(n)$ (a measure of patch occupancy) and total population size is $N_T = \Sigma np(n)$ (both summations are over n). Average local abundance is N_T/p_T.

Different populations coupled by dispersal comprise a metapopulation (Hanski, 1991). A rare species, in a geographical sense, has a small metapopulation ('spatial rarity'). In practice, the entire range of a species can rarely be usefully viewed as a single metapopulation, because of disjunct ranges and the sheer effects of isolation by distance in very large ranges (Maurer and Nott, in press).

As noted in the Introduction, when examining species over broad geographical areas there are several distinct facets of rarity, including total population size, N_T, and the number of local populations, p_T. As with local rarity, different dynamical reasons for spatial rarity may have different evolutionary consequences. One factor which arises in comparing spatially rare versus widespread species is that the latter may exhibit spatial heterogeneity in abundance and/or selective environments, which can have important implications for species-wide evolution.

For a fixed mean local abundance, N_T decreases linearly with decreasing p_T. A positive correlation between local abundance and range size (Brown, 1984; Lawton, 1993; Gaston, 1994) implies that spatially rare species have disproportionately low N_T, compared with this expectation. For neutral evolution, the evolutionarily relevant population size for species with stable dynamics is essentially N_T (Nei, 1987). Likewise, if the selective environment is spatially homogeneous with a single adaptive peak, the population size governing the input of potentially useful adaptive variation is ultimately the entire species' population. Our earlier conclusions about the effect (or lack thereof) of local population size on local evolution carry over wholesale to the effects of total population size on evolution at the level of entire species. This is basically Fisher's view (Wade, 1992).

As with local abundance, current range sizes may not accurately measure effective N_T. For instance, a species that experiences a sharp reduction in range, say because of climate change, may have a long transitional phase with considerably more genetic variation than expected from equilibrial genetic models.

As with local rarity, a species may be rare in a geographical sense for many different dynamical reasons. For instance, in a 'classical' meta-population (Hanski, 1991), the incidence of a species in a landscape is determined by a balance between colonization and extinction rates. Two species could be equally rare, one because of low colonization rates, the other because of high local extinction rates. They should evolve differently, despite their comparable range sizes; the species with frequent local extinction has little scope for evolution to hone local adaptations, compared with the species with low colonization rates.

One limiting case of spatiotemporal dynamics is to imagine that after speciation an initial bout of colonization spreads species over a landscape, with only minor subsequent dispersal among established populations (relative to local population sizes). Gene flow will not constrain local adaptation, but can still enrich local variation. The principal difference between spatially rare and widespread species will be in the number of populations making them up, scaling the pool of available variation, and defining the potential for a shifting balance process to operate.

There are hints in evolutionary theory that the size of a metapopulation (= number of local populations) may be important, independent of its effects on N_T. In the shifting balance process, any single population has a low probability of peak transition. The species-wide probability of a transition increases with p_T (Maynard Smith, 1983a; Newman et al., 1985; Holt, 1987; S. Wright, as reported in Moore and Tonsor, 1994). Hence, species with wide ranges (numerous demes, i.e. large p_T) might evolve more readily by a shifting balance mechanism than species with narrow ranges.

But there is a fundamental difference between the size of a local population and the size of a metapopulation (p_T). All individuals in a local population, to a reasonable approximation, experience the same environment. The different 'individuals' (= demes) of metapopulations necessarily occur in different places. The larger a metapopulation, the larger the area it encompasses. A basic fact about the earth's surface is that physical and biotic heterogeneity scale with area (Williamson, 1981). Such spatial heterogeneity within a species' range has several consequences:

- If it translates into spatial variation in the selective regime, different populations will have different local selective optima. This permits adaptive spatial differentiation, but vitiates the range size effect on the likelihood of a shifting balance process across an entire species, because different populations will be under the influence of different adaptive peaks. The literature on shifting balance has largely ignored spatial heterogeneity.
- There may be considerable variation in local abundance within species' ranges, a pattern which Lawton (1994) refers to as the spatial 'texture' of abundance. This may not affect evolution if the selective environment is spatially uniform. However, the details of range texture – and its

underlying dynamical causes – matter a great deal if selection varies spatially. With dispersal and interbreeding, selection within a species involves an averaging over space. The way this plays out is critically influenced by how the selective environment and local abundance covary.

For instance, Endler (1977) argued that gene flow did not prevent local selection from producing clinal variation. This conclusion was based on models in which populations at adjacent points along an environmental gradient had roughly equal abundances, leading to reciprocal effects of gene flow which cancelled out. By contrast, gene flow between populations varying in abundance may be highly asymmetrical. High rates of immigration into rare populations can prevent local adaptation (see above). If the spatial texture of abundance includes a great deal of local variation in abundance, rare populations will typically neighbour abundant populations, leading to potential high rates of gene flow into rare populations. By contrast, if abundances change gently along smooth environmental gradients, rare populations will mainly have other rare populations nearby; the rate of immigration may tend to scale proportionally with local carrying capacity. Thus, gene flow should have the greatest homogenizing effect if there is substantial yet spatially uncorrelated spatial variability in abundance.

If spatial variation in abundance and variability in the selective environment are correlated, there will be a bias across the species towards environments experienced by the greatest number of individuals. Recently, adaptive evolution in species with a source-sink structure (e.g. at a species' border: Parsons, 1991; Hoffman and Blows, 1994) has been analysed (Holt and Gaines, 1992; Holt, 1996a). Typically (though not always) natural selection is biased toward the maintenance and improvement of adaptation in the source, at the expense of improved adaptation in the sink.

One general implication is that an ecological factor restricting a species' range may indirectly foster the evolution of habitat specialization, making the range restriction evolutionarily stable. For instance, in source-sink models, at moderate to low rates of dispersal, as the sink becomes a less fit habitat, the intensity of selection for improving fitness there weakens (Holt, 1996). The reason is that the force of selection is blindly biased towards those environments actually experienced by individuals in which individuals contribute relatively most toward future generations. Individuals in sinks have low fitness, and there tend to be few of them, so adaptation to the sink is 'devalued' by selection, relative to adaptation to the source.

It has been argued (Holt and Gaines, 1992; Holt, 1996) that this asymmetry in selection can explain phylogenetic conservatism in niche properties. A tendency towards niche conservatism may be greatly enhanced if rare populations experience extinction, and are re-colonized from persistent, abundant populations (Harrison, 1991). Internal range dynamics can greatly influence a species' effective N_T. Whitlock (ms) shows that in metapopulations comprised of sources and sinks, effective N_e can be reduced far below census N_T. An understanding of the temporal dynamics of species' ranges is thus required to gauge the relative potential for long-term evolution in localized versus widespread species.

An important direction for future work will be to draw out the implications of spatial texture, spatial dynamics and range size for the rates of generation of evolutionary novelties, and speciation. Here are some closing thoughts.

- As noted above for within-community evolution, if the spatial texture of abundance is such that there are many more rare than common populations in a species' range, significant niche evolution within a lineage might be associated with rare populations, simply because there are more of them. However, this is not likely if most rare populations are sink populations, or experience rapid rates of extinction with re-colonization from more persistent, common populations (as in the island model above).

- Spatially restricted species may be more likely to evolve specific adaptations to local conditions (Futuyma, 1989). Such adaptations may be ephemeral in more widespread species, whose phenotypes express adaptive compromises to a heterogeneous, ever-shifting environment (Futuyma, 1989; Rosenzweig, 1995).

- Rosenzweig (1995) argues that the likelihood of allopatric speciation (number of daughter species per ancestral species) increases with range size, at least for small to medium ranges. There are two distinct effects here: the likelihood of isolate formation increases with range size, and large ranges may encompass a greater range of selective environments, permitting more divergent evolution among populations. This suggests that, as Darwin believed, 'common species' (= widespread species) should be the fount of speciation – even if speciation itself occurs most frequently in rare populations of those widespread species.

- Sexual species may have a cost of rarity, for instance because individuals must allocate more energy and time to finding mates. This cost may be most manifest in competitive interactions with less rare species, which do not incur such costs and thus can be more efficient resource exploiters. Hopf and Hopf (1985) and Michod (1995, Chapter 10) have argued that this effect of rarity may have important macroevolutionary conse-quences. Rare sexual species may be differentially vulnerable to competitive exclusion, leading to gaps in species' utilization functions along resource gradients. This mechanism works most forcefully when 'rarity' denotes low local population abundance, rather than range size. Community studies often reveal that habitat partitioning is far more evident and fine-grained than local within-site resource partitioning among related taxa (Schoener, 1974). An intriguing possibility is that this generalization about community patterns may indirectly reflect the role of rarity in evolution.

We are just beginning to fathom the implications of spatial dynamics and spatial textures in abundance for evolution.

12.4 CONCLUSIONS

The problem of analysing the implications of rarity (viz., low population size and/or small range size) for evolution upon inspection has resolved into an interwoven tangle of problems. To reiterate this chapter's opening remarks, sensible questions about rarity and evolution require one to have a clear comparison in mind. Beyond this rather obvious (but often forgotten) point, the ruminations presented above are woven together by an emphasis on the importance of basic population dynamics. To reprise some of the main points:

- The effective population size of a species may differ greatly from its census size, because of long-term fluctuations in local abundance, range size, local extinctions and source-sink dynamics.
- The selective consequences of rarity cannot be evaluated without a crisp understanding of the dynamical reasons for rarity. Often, selection will not act in such a way as to make a rare species less rare; indeed, making a factor more important in limiting population size, or geographical range, may sometimes make it less important in selection. Ecological factors can affect population size with no effects on selection; little is known about how spatial variation in abundance reflecting ecological variation maps on to spatial variation in the selective environment.
- Considering the demographic effects of immigration into a sink reveals that the constraint on local adaptation may not quite be that gene flow hampers local selection (the conventional wisdom), but simply that the immigrants are maladapted to the local environment to start with. Ignoring the mechanisms of population dynamics can lead to a misunderstanding of the causal factors in evolutionary dynamics.

A clearer understanding of how population abundance and range size act as ecological drivers for evolutionary dynamics is particularly important for addressing many applied ecological problems. For instance, there is increasing focus on the genetics of rare populations, motivated by conservation concerns (e.g. Lande 1988; Falk and Holsinger, 1991; Schemske *et al.*, 1994). If population size scales the potential evolutionary response of species to changing environments, a species which has become rare because of environmental indignity A (e.g. habitat destruction) may thereby be unable to adapt to further environmental indignities X, Y and Z (e.g. toxic wastes, pressure from invading exotics). As another example, the entire point of pest control programmes is to turn common species into rare species. For control to be anything other than a short-term palliative, evolutionary dynamics of the target pest must be considered. Evolutionarily stable pest control requires the imposition of strong limiting factors that do not evoke strong countervailing selection in the pest. Some of the most significant problems facing our species today thus lie squarely on the cusp between the ecological causes and evolutionary consequences of rarity.

REFERENCES

Aarssen, L.W. (1983) Ecological combining ability and competitive combining ability in plants: towards a general evolutionary theory of coexistence in systems of competition. *The American Naturalist*, **122**, 707–731.

Abrams, P. (1986) Adaptive responses of predators to prey and prey to predators: the failure of the arms race analogy. *Evolution*, **40**, 1229–1247.

Antonovics, J. (1976) The nature of limits to natural selection. *Annals of the Missouri Botanical Garden*, **63**, 224–248.

Arnold, A.J. and Fristrup, V. (1982) The theory of evolution by natural selection: a hierarchical expansion. *Paleobiology*, **8**, 113–129.

Barrett, S.C.H. and Kohn, J.R. (1991) Genetic and evolutionary consequences of small population size in plants: implications for conservation, in *Genetics and Conservation of Rare Plants* (eds Falk and Holsinger), Oxford University Press, Oxford, pp. 3–30.

Barton, N.H. (1989) Founder effect speciation, in *Speciation and its Consequences* (eds D. Otte and J. Endler), Sinauer Associates, Sunderland, MA, pp. 229–256.

Barton, N.H. and Rouhani, S. (1992) Adaptation and the shifting balance. *Genetical Research*, **61**, 57–74.

Begon, M. (1992) Density and frequency dependence in ecology: messages for genetics? in *Genes in Ecology* (eds R.J. Berry, T.J. Crawford, and G.M. Hewitt), Blackwell, London, pp. 335–352.

Brown, J.H. (1984) On the relationship between distribution and abundance. *The American Naturalist*, **124**, 255–279.

Brown, J.H. and Kodric-Brown, A. (1977) Turnover rates in insular biogeography: effect of immigration on extinction. *Ecology*, **58**, 445–449.

Charlesworth, B. (1994) *Evolution in Age-structured Populations*, 2nd edn, Cambridge University Press, Cambridge.

Charlesworth, B., Lande, R. and Slatkin, M. (1982) A new-Darwinian commentary on macroevolution. *Evolution*, **36**, 474–498.

Coyne, J.A. (1994) Ernst Mayr and the origin of species. *Evolution*, **48**, 19–30.

Crawford, T.J. (1984) What is a population? in *Evolutionary Ecology* (ed B. Shorrocks), Blackwell Scientific, Oxford, pp. 135–174.

Crow, J.F., Engels, W.R. and Denniston, C. (1990) Phase three of Wright's sifting-balance theory. *Evolution*, **44**, 233–247.

Darwin, C. (1859 [1964]) *On the Origin of Species, A Facsimile of the First Edition*, Harvard University Press, Cambridge, MA.

Diamond, J.M. (1984) 'Normal' extinctions of isolated populations, in *Extinctions* (ed. M.H. Nitecki), University of Chicago Press, Chicago, pp. 191–246.

Eldredge, N. (1985) *Unfinished Synthesis: Biological Hierarchies and Modern Evolutionary Thought*, Oxford University Press, Oxford.

Eldredge, N. and Gould, S.J. (1988) Punctuated equilibrium prevails. *Nature*, **332**, 211–212.

Endler, J. (1977) *Geographic Variation, Speciation, and Clines*, Princeton University Press, Princeton, NJ.

Falk, D.A. and Holsinger, K.E. (1991) *Genetics and Conservation of Rare Plants*, Oxford University Press, Oxford.

Futuyma, D.J. (1989) Macroevolutionary consequences of speciation: inferences from phytophagous insects, in *Speciation and Its Consequences* (eds D. Otte and J.A. Endler), Sinauer Associates, Inc., Sunderland, MA, pp. 557–578.

Gaston, K.J. (1994) *Rarity*, Chapman & Hall, London.

Gillespie, J.H. (1991) *The Causes of Molecular Evolution*, Oxford University Press, Oxford.

Gomulkiewicz, R. and Holt, R.D. (1995) When does evolution by natural selection prevent extinction? *Evolution*, **49**, 201–207.

Grant, B.R. and Grant, P.R. (1989) *Evolutionary Dynamics of a Natural Population*, University of Chicago Press, Chicago.

Grant, V. (1981) *Plant Speciation*, 2nd edn, Columbia University Press, New York.

Hamrick, J.L., Godt, M.J.W., Murawski, D.A. and Loveless, M.D. (1991) Correlations between species traits and allozyme diversity: implications for conservation biology, in *Genetics and Conservation of Rare Plants* (eds D.A. Falk and K.E. Holsinger), Oxford University Press, Oxford, pp. 75–86.

Hanski, I. (1991) Single-species metapopulation dynamics: concepts, models, and observations. *Metapopulation Dynamics* (eds M. Gilpin and I. Hanski), Academic Press, New York, pp. 17–38.

Harrison, S. (1991) Local extinction in a metapopulation context: an empirical evaluation. *Metapopulation Dynamics* (eds M. Gilpin and I. Hanski), Academic Press, New York, pp. 73–88.

Hastings, A. and Harrison, S. (1994) Metapopulation dynamics and genetics. *Annual Review of Ecology and Systematics*, **25**, 167–188.

Hedrick, P. (1983) *Genetics of Populations*, Van Nostrand International, New York.

Hill, W.G. and Caballero. A. (1992) Artificial selection experiments. *Annual Review of Ecology and Systematics*, **23**, 287–310.

Hoffmann, A.A. and Blows, M.W. (1994) Species borders: ecological and evolutionary perspectives. *Trends in Ecology and Evolution*, **9**, 223–227.

Holt, R.D. (1977) Predation, apparent competition, and the structure of prey communities. *Theoretical Population Biology*, **12**, 197–229.

Holt, R.D. (1987) Population dynamics and evolutionary processes: the manifold effects of habitat selection. *Evolutionary Ecology*, **1**, 331–347.

Holt, R.D. (1990) The microevolutionary consequences of climate change. *Trends in Ecology and Evolution*, **5**, 311–315.

Holt, R.D. (1993) Ecology at the mesoscale: The influence of regional processes on local communities, in *Species Diversity in Ecologyical Communities* (eds R. Ricklefs and D. Schluter), University of Chicago Press, Chicago, pp. 77–88.

Holt, R.D. (1995) Linking species and ecosystems: Where's Darwin? in *Linking Species and Ecosystems* (eds C.G. Jones and J.H. Lawton), Chapman & Hall, New York, pp. 273–279.

Holt, R.D. (1996a) Adaptive evolution in source-sink environments: Direct and indirect effects of density-dependence. *Oikos*, **75**, 182–192.

Holt, R.D. (1996b) Demographic constraints in evolution: unifying the evolutionary theories of senescence and niche conservatism. *Evolutionary Ecology*, **10**, 1–11.

Holt, R.D. and Gaines, M.S. (1992) Analysis of adaptation in heterogeneous landscapes: implications for the evolution of fundamental niches. *Evolutionary Ecology*, **6**, 433–447.

Holt, R.D. and Gomulkiewicz, R. (in press) The influence of immigration on local adaptation: reexamining a familiar paradigm. *The American Naturalist*.

Hopf, F.A. and Hopf, F.W. (1985) The role of the Allee effect in species packing. *Theoretical Population Biology*, **27**, 27–50.

Husband, B.C. and Barrett, S.C.H. (1992) Effective population size and genetic drift in tristylous *Eichornia paniculata* (Pontederiaceaea). *Evolution*, **46**, 1875–1890.

Kimura, M. (1983) *The Neutral Theory of Molecular Evolution*, Cambridge University Press, Cambridge, UK.

Kirkpatrick, M. (1982) Quantum evolution and punctuated equilibria in continuous genetic characters. *The American Naturalist*, **119**, 833–848.

Kunin, W.E. and Gaston, K.J. (1993) The biology of rarity: patterns, causes, and consequences. *Trends in Ecology and Evolution*, **8**, 298–301.

Lande, R. (1979) Effective deme sizes during long-term evolution estimated from rates of chromosomal rearrangement. *Evolution*, **33**, 234–251.

Lande, R. (1980) Genetic variation and phenotypic evolution during allopatric speciation. *The American Naturalist*, **117**, 463–479.

Lande, R. (1988) Genetics and demography in biological conservation. *Science*, **241**, 1455–1450.

Lande, R. (1994) Risk of population extinction from fixation of new deleterious mutations. *Evolution*, **48**, 1460–1469.

Lawton, J.H. (1993) Range, population abundance and conservation. *Trends in Ecology and Evolution*, **8**, 409–413.

Lawton, J.H. (1994) Population dynamic principles. *Philosophical Transactions of the Royal Society of London, B*, **344**, 61–68.

Leigh, E.G. Jr (1986) Ronald Fisher and the development of evolutionary theory. I. The role of selection. *Oxford Surveys in Evolutionary Biology*, **3**, 187–223.

Levins, R. (1979) Coexistence in a variable environment. *The American Naturalist*, **114**, 765–783.

Lynch, M. (1994) Neutral models of phenotypic evolution, in *Ecological Genetics* (ed. L. Real), Princeton University Press, Princeton, pp. 86–108.

Lynch, M. and Gabriel, W. (1990) Mutation load and the survival of small populations. *Evolution*, **44**, 1725–1737.

Lynch, M. and Lande, R. (1993) Evolution and extinction in response to environmental change, in *Biotic Interactions and Global Change* (eds P.M. Kingslover and R.B. Huey), Sinauer Associates, Inc. Sunderland, Massachusetts, pp. 234–250.

MacArthur, R.H. (1972) *Geographical Ecology: Patterns in the Distribution of Species*, Harper and Row, New York.

MacArthur, R.H. and Wilson, E.O. (1967) *The Theory of Island Biogeography*, Princeton University Press, Princeton.

Maurer, B.A. and Nott, M.P. (in press) Geographic range fragmentation and the evolution of biological diversity, in *Biodiversity Dynamics* (eds M.L. McKinney and J.A. Drake), Columbia University Press.

May, R.M. (1975) Patterns of species abundance and diversity, in *Ecology and Evolution of Communities* (eds M.L. Cody and J.M. Diamond), Harvard University Press, Cambridge, MA, pp. 81–120.

Maynard Smith, J. (1983a) Current Controversies in Evolutionary Biology, in *Dimensions of Darwinism* (ed. M. Grene), Cambridge University Press, Cambridge, pp. 273–278.

Maynard Smith, J. (1983b) The genetics of stasis and punctuation. *Annual Review of Genetics*, **17**, 11–25.

Maynard Smith, J. (1989) *Evolutionary Genetics*, Oxford University Press, Oxford.

Mayr, E. (1982) *The Growth of Biological Thought*, Harvard University Press, Cambridge, MA.

McCauley, D.E. (1993) Evolution in metapopulations with frequent local extinction and recolonization, in *Oxford Surveys in Evolutionary Biology*, Vol. 9 (eds D. Futuyma and J. Antonovics), Oxford University Press, Oxford, pp. 109–134.

Merrell, D.J. (1994) *The Adaptive Seascape: The Mechanism of Evolution*, University of Minnesota Press, Minneapolis, MN.

Michod, R.E. (1995) *Eros and Evolution: A Natural Philosophy of Sex*, Addison-Wesley, Reading, MA.

Moore, F.B.-G. and Tonsor, S.J. (1994) A simulation of Wright's shifting-balance process: migration and the three phases. *Evolution*, **48**, 69–80.

Nagylaki, T. (1979) The island model with stochastic migration. *Genetics*, **91**, 163–176.

Nei, M. (1987) *Molecular Evolutionary Genetics*, Columbia University Press, New York.

Nei, M. and Grauer, D. (1984) Extent of protein polymorphism and the neutral mutation theory. *Evolutionary Biology*, **17**, 73–118.

Newman, C.M., Cohen, J.E. and Kipnis, C. (1985) Neo-darwinian evolution implies punctuated equilibrium. *Nature*, **315**, 400–401.

Olivieri, I., Couvet, D. and Gouyon, P. (1990) The genetics of transient populations: research at the metapopulation level. *Trends in Ecology and Evolution*, **5**, 207–209.

Parsons, P.A. (1991) Evolutionary rates: stress and species boundaries. *Annual Review of Ecology and Systematics*, **22**, 1–18.

Pimentel, D., Feinberg, E.H., Wood, P.W. and Hayes, J.T. (1965) Selection, spatial distribution, and the coexistence of competing fly species. *The American Naturalist*, **99**, 97–109.

Pimm, S.L. (1991) *The Balance of Nature?* University of Chicago Press, Chicago.

Provine, W.B. (1989) Founder effects and genetic revolutions in microevolution and speciation: an historical perspective, in *Genetics, Speciation and the Founder Principle* (eds L.V. Giddings, K.Y. Kaneshiro and W.W. Anderson), Oxford University Press, Oxford, pp. 43–76.

Pulliam, H.R. (1988) Sources, sinks, and population regulation. *The American Naturalist*, **132**, 652–661.

Ridley, M. (1993) *Evolution*, Blackwell, London.

Rosenzweig, M.L. (1995) *Species Diversity in Space and Time*, Cambridge University Press, Cambridge.

Rosenzweig, M.L., Brown, J.L. and Vincent, T.L. (1987) Red Queens and ESS: the coevolution of evolutionary rates. *Evolutionary Ecology*, **1**, 59–94.

Schaffer, W.M. and Rosenzweig, M.L. (1978) Homage to the Red Queen. I. Coevolution of predators and their victims. *Theoretical Population Biology*, **14**, 135–157.

Schemske, D.W., Husband, B.C., Ruckelshaus, M.H. *et al.* (1994) Evaluating approaches to the conservation of rare and endangered plants. *Ecology*, **75**, 584–606.

Schoener, T.W. (1974) Resource partitioning in ecological communities. *Science*, **185**, 27–39.

Schoener, T.W. and Spiller, D.A. (1987) High population persistence in a system with high turnover. *Nature*, **330**, 474–477.

Seger, J. (1992) Evolution of exploiter–victim relationships, in *Natural Enemies: The Population Biology of Predators, Parasites and Diseases* (ed. M. Crawley), Blackwell, Oxford, pp. 3–25.

Stanley, S.M. (1990) The general correlation between rate of speciation and rate of extinction: fortuitous causal linkages, in *Rates of Evolution* (ed. R.M. Ross), University of Chicago Press, Chicago, pp. 103–127.

Tokeshi, M. (1993) Species abundance patterns and community structure. *Advances in Ecological Research*, **24**, 112–186.

Wade, M.J. (1992) Sewall Wright: gene interaction and the shifting balance theory, in *Oxford Surveys in Evolutionary Biology*, Vol. 8 (eds D. Futuyma and J. Antonovics), Oxford University Press, Oxford.

Whitlock, M. (ms.) Sources and sinks: asymmetric migration, population structure, and the effective population size.

Williamson, M. (1981) *Island Populations*, Oxford University Press, Oxford.

Wilson, D.S. (1989) The diversification of single gene pools by density- and frequency-dependent selection, in *Speciation and its Consequences* (eds D. Otte and J.A. Endler), Sinauer Associates, Inc., Sunderland, MA, pp. 366–385.

Wright, S. (1931) Evolution in Mendelian populations. *Genetics*, **16**, 97–159.

Wright, S. (1977) *Evolution and the Genetics of Populations. Vol. 3 Experimental Results and Evolutionary Deductions; Vol. 4 Variability within and among Natural Populations*. University of Chicago Press, Chicago.

PART THREE: FUTURE DIRECTIONS

13 Predicting and understanding rarity: the comparative approach

Peter Cotgreave and Mark Pagel

Over the next several decades, actions will need to be taken towards preserving the many species now facing extinction. Unfortunately, this will be done in the context of enormous ignorance about most of these species.

G. Mace, 1995

13.1 INTRODUCTION

Our ignorance about the world's organisms is astonishing. For many species, we do not even know whether they are rare or common. For others, the fact that they are rare is the most significant piece of information we have. In a small number of cases, we know how rare a species is and also have a good knowledge of its life history, ecology or behaviour. When we have this information about a whole group of species, we can ask what categories of species tend to be rare, merely by tallying the characteristics of known rare species. Such tallies have their uses but, for two reasons, they may not necessarily reveal what features of the species are the cause of their rarity. First, variables that are correlated with rarity could equally well be causes or consequences of that rarity and we must be careful in deciding between these alternatives. Second, correlations found between two or more variables that are measured across species may be coincidental artefacts of the methods used and not true evolutionary correlations, by which we mean that the two characters have evolved together on a number of independent occasions. The first problem is undoubtedly complex and difficult but it can occasionally be overcome. Towards the end of the chapter we will demonstrate how. However, this chapter is mainly about solving the second problem – how to identify evolutionary correlations.

The Biology of Rarity. Edited by William E. Kunin and Kevin J. Gaston.
Published in 1997 by Chapman & Hall, London. ISBN 0 412 63380 9.

The techniques of studying evolutionary correlations across species are those of the comparative method (Ridley, 1983; Pagel and Harvey, 1988; Harvey and Pagel, 1991) and we shall explore a variety of these methods in this chapter. We begin with some simple, exploratory statistics, useful for making predictions about rarity, before considering newer, more sophisticated methods for recognizing ways in which rarity has co-varied with other features over evolutionary time. We use different measures of rarity – range size, population density, endangeredness – in different examples, merely to demonstrate the versatility of the methods.

13.2 THE BIRDS OF AUSTRALIA

To illustrate the use and limitations of comparative techniques in understanding rarity we have collated data on the 559 species of non-coastal birds in Australia (Figure 13.1). The variables in this data set include measures of body size, diet and breeding characteristics, as well as data on population density and geographical range size.

Population density and geographical range size are two commonly employed measures of rarity and it is not unusual to see them used interchangeably. One striking feature of the data for our birds is that regardless of which measure is used, most species are rare (Figure 13.2). However, among the birds of Australia, these two measures are slightly negatively correlated: widespread species tend to live at low population densities (Figure 13.3). Already then, rarity among this group is revealed to be a complex phenomenon, and understanding variation in one type of rarity may not shed any light at all on another.

In fact, the relationship shown in Figure 13.3 may be misleading because, unsurprisingly, widespread species are more likely to have been studied than geographically restricted ones. The subset of species included in Figure 13.3 – those for which we have data on both geographical range size and population density – have an average range size of 224 squares of the grid used by Blakers *et al.* (1984), whereas the average range size for those not included (because there is no published measure of population density) is 151 squares. Because the subset of species for which data on population density exist is a biased sample, and because small geographical range size is commonly associated with becoming threatened, our illustrations of how to investigate rarity will concentrate on geographical range size.

We raise the issue of bias early on because it plagues lists of threatened species. Mace (1995) estimates, for example, that only about 50% of mammal species and perhaps 5% of fish have been investigated with respect to their threatened status, and invertebrates are even less well studied. Vigilance about sources of bias is essential in comparative studies of rarity.

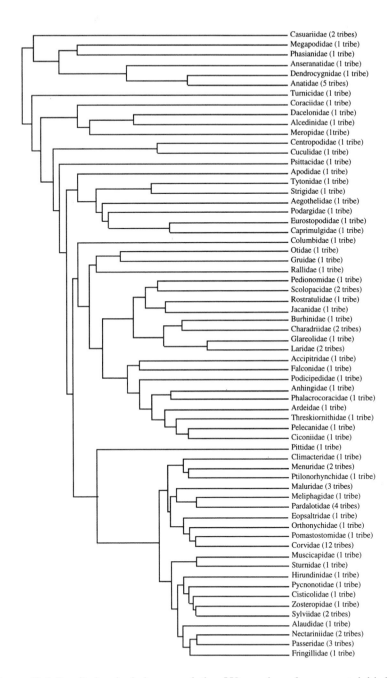

Figure 13.1 Family-level phylogeny of the 559 species of non-coastal birds of Australia (from Sibley and Ahlquist, 1990). For the analyses, the whole phylogeny was used, including the branching structure below the family level.

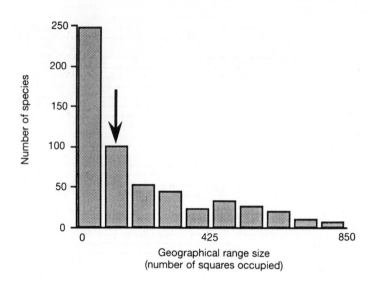

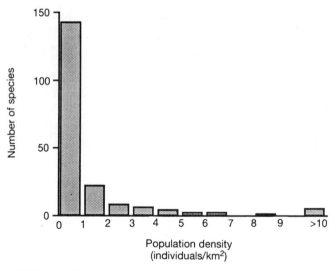

Figure 13.2 Distribution of geographical range sizes and population densities among the birds of Australia (Blakers *et al.*, 1984). Note that most species are rare, whichever of the two measures is used. The plot for range size includes all 559 species, while population density data were available for only 192 species. The arrow on the range size plot indicates a range of 100 squares and we have used this as the arbitrary boundary between rare and common species.

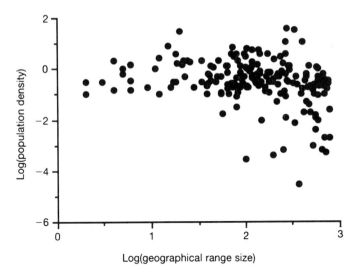

Figure 13.3 Population density and geographical range size are negatively correlated (on logarithmic axes, $n = 192$, $r = 0.27$, $P = 0.0001$). However, this relationship may be confounded by the fact that population density data tend to be available only for widespread species.

13.3 PREDICTIVE ANALYSES: IDENTIFYING THE CHARACTERISTICS OF RARE SPECIES

13.3.1 What sorts of birds are rare?

The simplest way to begin to investigate factors associated with rarity is to compare the characteristics of rare species with those of species that are not rare. However, even for such simple comparisons, conventional statistics, which treat each species as an independent data point, will overestimate the true number of degrees of freedom (this point will be developed later). Roughly speaking, for species drawn from a bifurcating phylogeny, the actual degrees of freedom in this sort of test will be approximately half of the number of species in each group. Even then, the individual species values cannot be assumed to have equal variances, as the null hypothesis assumes (for reasons that are explained by Harvey and Pagel, 1991). Making tallies is easy but their interpretation is less so. Accordingly, we do not report any statistical probabilities in this section.

We have defined rare species as those with a geographical range size of less than 100 of the blocks used by Blakers *et al.* (1984) (Figure 13.2), and compared a number of their characteristics with the characteristics of the remaining species (Figure 13.4). Rare species have higher population densities, are smaller, have been less well studied, have fewer known

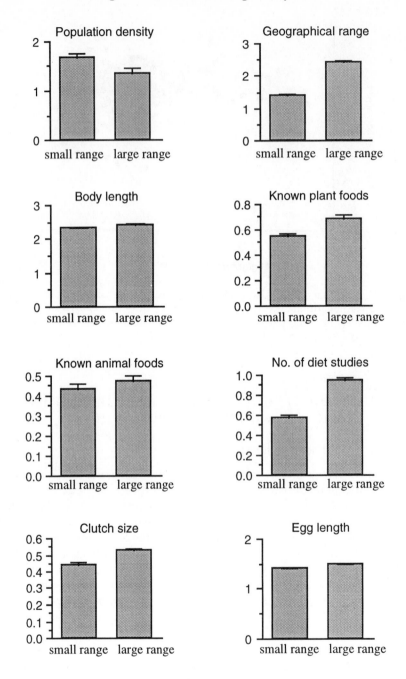

Figure 13.4 Comparisons of variables used in later analyses between rare species and those that are not rare (means of logged data ± SE; some of the standard error bars are too small to be seen). Rare species are defined as occupying fewer than 100 of the 850 or so blocks of the maps in Blakers *et al.* (1984).

animal and plant foods and have small eggs and small clutches. One of the most striking differences between rare and not-rare species concerns the number of published studies about a species' diet. This simply reflects the fact that widespread species are more likely to be studied but it may turn out to be important to control for such artefacts in later analyses. It is also important to observe that rarity is relative – there is no unique definition of a rare species. Our results might have been different if we had chosen 120 blocks as the cut-off point rather than 100. Equally, we must be careful to observe that no generalizations about other taxa can be made from analyses of this type. What is true of Australian birds may or may not be true of other groups.

These sorts of predictive analyses need not be limited to contemporary features of a species. We can, for example, begin to investigate questions related to phylogeny, and hence evolution. For example, in the Australian birds, non-passerine species have geographical ranges that are, on average, nearly one and a half times the size of the ranges of passerine species. This is a specific case of comparing one clade with all other birds; we could equally well look more generally at features of particular groups of organisms.

Another sort of phylogenetic comparison looks not across different groups but backwards through time. Nee *et al.* (1991), Cotgreave and Harvey (1991, 1994) and Cotgreave (1994) have compared population densities in 'ancient' and 'young' tribes. We now compare such groups to see whether they differ in how likely their species are to be endangered. Tribal age is measured using Sibley and Ahlquist's (1990) arbitrary units of distance along the branches of their DNA–DNA hybridization phylogeny of the birds. Values range from 3.3 (young tribes) to 27.0 (old tribes) but the distribution is bimodal; no tribe falls in the range 17.0 to 18.0. We therefore used this as the boundary between ancient and young tribes and calculated that the probability of being officially 'threatened' (IUCN, 1986) is 4% for species in young tribes and 12% for species in ancient tribes. This hints at a strong link between rarity and phylogeny, which may help to unravel the evolutionary reasons why some species are rare.

It is important to bear in mind that none of the results reported in this section necessarily tells us anything about the evolutionary causes of rarity. The value of the results is that they can tell us what sorts of species are shouldering more of the 'threat load'.

13.4 EXPLANATORY ANALYSES: IDENTIFYING EVOLUTIONARY CORRELATES OF RARITY

13.4.1 Methods for the analysis of continuously varying characters

The problem with the kind of statistics presented in the previous section is that they are no more than descriptive summaries. They can hint at answers but they cannot explain what causes rarity. Although the true reasons why species are rare will always be elusive, it could be helpful to look for evolutionary correlates of the various types of rarity. Here the search is for the evolutionary changes that may 'unwittingly' put a species at risk of becoming rare. This may help to identify a profile of a hypothetical species most likely to be (or to become) rare and, importantly, may help us to understand why some evolutionary pathways seem to lead to rarity. Such an understanding could be crucial in terms of conservation.

The comparative approach recognizes that treating each separate species as being evolutionarily independent of all others, as do the predictive analyses, may produce misleading answers (Ridley, 1983; Pagel and Harvey, 1988; Harvey and Pagel, 1991). Two or more species that are closely related will tend to be more similar to one another than are two unrelated species. This means that species are not independent for statistical purposes and to count them as such when calculating probabilities will tend to overestimate the strength of a result. Lack of independence among species is not confined to morphological traits. Behavioural traits, or traits that arise from behavioural patterns, such as geographical range size or population density cannot be assumed to be independent among species. It is not intuitively obvious that these population and species level characteristics should be correlated with phylogeny in the way that, for example, morphology is. Nevertheless, such phylogenetic correlations do exist (e.g. Lawton, 1993). Related species have similar values of population density (Cotgreave and Harvey, 1992) and among birds in our data set there is generally more variation in geographical range size among different taxonomic orders of birds than there is within them (Figure 13.5).

To solve the problem of non-independence of species, several statistical techniques have been developed that exploit the phylogenetic relationships among species to extract independent pieces of information. These independent data points are then used in standard statistical analyses.

13.4.2 The method of independent contrasts

One of the most understandable and widely used methods for analysing comparative data was described by Felsenstein (1985) and can be illustrated by reference to the example given in Figure 13.6. The horizontal arrows (d1, d2, d3) show that we begin by simply finding the difference between pairs of extant species, for example the finch and riflebird, on the traits of interest. The difference between any pair of species represents an amount of evolutionary divergence that has accumulated since they speciated from their common ancestor. The phylogeny in Figure 13.6 permits three comparisons between extant species. Next we make comparisons between internal nodes of the phylogeny (d5, d6), or in some instances between a species and a node (d4). The arrows indicate the six comparisons that this phylogeny uniquely defines – a strength of this

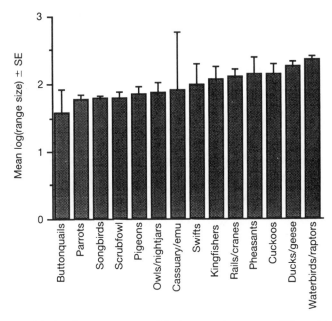

Figure 13.5 Variation in mean geographical range size among different taxonomic orders of birds in Australia. There is more variance among than within orders ($F_{13.543} = 5.645$, $P = 0.0001$).

approach to which we shall return. In general, a bifurcating phylogeny with n species yields $(n - 1)$ such comparisons, corresponding to the $(n - 1)$ degrees of freedom in the data. The important point is that each of these 'contrasts' or differences is independent of all of the others. The evolution in each branch of the tree has only been counted once.

In reality, we do not know what the ancestral species at the nodes were like, and the value placed at the internal node is not technically an attempt to estimate the past. Rather the values at each node will be some weighted average of the two species (or nodes) that evolved from it, with weights that are usually related to the branch lengths. This means that the values at the nodes are chosen so that the comparisons are mutually independent, although how precisely to achieve this in all cases is controversial (Felsenstein, 1985; Harvey and Pagel, 1991; Grafen, 1992; Pagel, 1992).

Once a set of contrasts has been found for two or more variables they can be plotted as in the graph inset in Figure 13.6. Their correlation and regression (forced through the origin) have exactly the same interpretation as the same statistics calculated across species (Pagel, 1993). In fact, the slope of the line relating two sets of contrasts estimates the relationship between changes in the two variables that occurred whilst they evolved along the branches of the phylogenetic tree. This additionally means that it

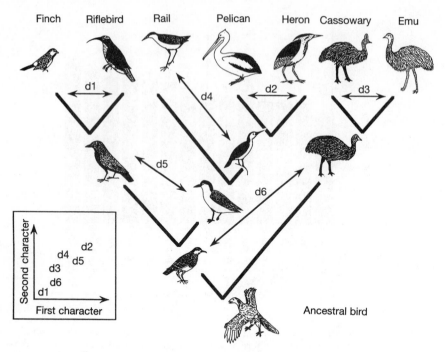

Figure 13.6 Phylogeny of seven species of Australian birds. When looking for an evolutionary correlation between two continuously varying characters, we calculate a comparison, or contrast, across each bifurcation on the tree. Thus, we compare the riflebird with the finch as one contrast (d1) and calculate another contrast between the common ancestor of these two species and the common ancestor of the rail, heron and pelican (d5). Contrasts are plotted as in the inset graph to test evolutionary hypotheses. The ancestors on this figure are based on imagination; in reality, we do not know what they were like, so we estimate their characteristics from the extant species.

is not necessary to use so-called directional comparative methods for continuous variables (methods designed to detect the correlation of changes along the branches of the phylogeny) because they measure the same changes as the independent contrast techniques (Pagel, 1993).

(a) An example

For the purpose of illustrating the method of independent contrasts, we will use geographical range size as the measure of rarity, ignoring other possibilities. Population density has been analysed elsewhere for the Australian avifauna (Cotgreave, 1995). As likely correlates of rarity (Figure 13.4), we will analyse body size as well as the breeding and diet of the species. The typical number of eggs in a clutch and the average length of an egg (Anon., 1990) have been used as very rough summary measures of life history strategy. Dietary information was taken from Barker and Vestjens (1990) and we have recorded the number of different animal foods and the number of different plant foods known to be eaten by each species. To control for sampling effort, we have also recorded the number of different published studies used to compile the feeding data on each species (including Barker and Vestjens' own unpublished records as one study).

All of these data are continuously varying, so it was possible to calculate a set of evolutionary contrasts for each variable (like those in Figure 13.6), after logarithmic transformation to improve the normality of the distributions. The contrasts were calculated using Purvis's (1992) package of software for the implementation of Pagel's (1992) modification of the original method (Felsenstein, 1985). Although it is contentious, Sibley and Ahlquist's (1990) phylogeny is the most objective phylogeny of the birds to date (Cotgreave and Harvey, 1994; Mooers and Cotgreave, 1994) and we have consequently used it in our analyses (Figure 13.1).

The contrasts in geographical range size were regressed on the contrasts in the predictor variables, in a multiple regression, and the results are presented in Table 13.1. The results in the table control simultaneously for all other variables. Body length, the number of plant foods and the size of the eggs all fail to correlate with geographical range size, after removing the effects of the other variables. However, clutch size is positively

Table 13.1 Multiple regression analyses of factors correlated with geographical range size in Australian non-coastal birds

Independent variable	Evolutionary contrasts method		Across-species analysis	
	Increase in total variance explained when variable is entered last	P	Increase in total variance explained when variable is entered last	P
log(body length)	0.3%	0.23	0.0%	0.67
log(number of known plant foods + 1)	0.1%	0.83	0.5%	0.02
log(number of known animal foods + 1)	2.1%	0.001	4.5%	< 0.0001
log(number of food studies)	15.2%	< 0.0001	16.0%	< 0.0001
log(egg length)	0.3%	0.23	0.0%	0.44
log(mean clutch size)	1.1%	0.02	2.9%	< 0.0001

For the analysis of evolutionary contrasts, it is necessary to force the regression through the origin. Three of the six independent variables explain significant amounts of the variation in range size ($F_{6,231} = 55.5$, $P = 0.0001$) and the full model explains a total of 59.1% of the variation. For the cross-species multiple regression, $F_{6,531} = 96.8$, $P = 0.0001$. Four of the variables explain significant amounts of the variance in log(range size). Analyses of the second type are inappropriate for the study of evolutionary relationships because species data are known to be distributed non-randomly with respect to phylogeny.

correlated with range size, suggesting perhaps that birds with a particular type of life history strategy are distributed widely, while other strategies lead to more confined distributions. The number of published studies about a species' diet is overwhelmingly the strongest correlate of geographical range size (Figure 13.7), accounting for an additional 15% of the variance after controlling for other effects – an effect 7.5 times greater than any other predictor. This probably reflects the fact that widely distributed species have been well studied generally and underscores our admonition that bias is potentially a common problem in studies of rarity.

The number of different animal foods eaten is independently and positively associated with geographical range size. Perhaps species with more types of animal food are more generalist or simply more carnivorous. However, if generalism were important, we should expect changes in the number of plant foods eaten to be associated with geographical range size. Table 13.1 shows that this is not the case.

(b) Independent contrasts versus correlations across species

In our data, all of the variables that show significant phylogenetic associations with geographical range size (based on contrasts) also show

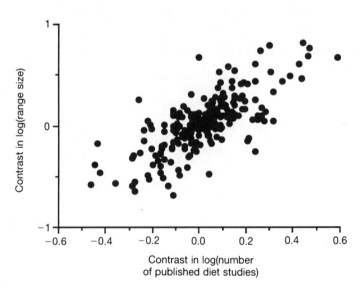

Figure 13.7 There is an 'evolutionary' relationship between the number of published studies about birds' diets and their geographical range sizes. Estimating the 'ancestral' number of papers published at a node on the phylogeny is logically nonsense. Nevertheless, the number of published studies may be similar for related species. Thus, it is clear that we should control for this pattern in our data before attempting to test for true evolutionary relationships between number of known food types and range size.

significant relationships across species, when included in a multiple regression (Table 13.1). However, although not significant, the across-species correlation between body length and range size is in the opposite direction to the correlation in independent contrasts. Further, across species, the number of plant foods eaten by a species is also a significant predictor of range size. This is the usual pattern of difference between analyses based on independent contrasts and those based on species: some significant results in the species analyses may disappear when the data are analysed using a technique that takes into account the non-independence of species.

(c) The requirement for phylogenetic information

In theory, the independent contrast approach requires a fully resolved phylogeny. That is one reason why we chose birds as an example group; there is no other taxon of comparable size for which an objective phylogeny is available (Mooers and Cotgreave, 1994). A more likely scenario, however, is that only a partially resolved phylogeny, or even a taxonomy, will be available. Fortunately, two methods (Grafen, 1989; Pagel, 1992) exist for calculating independent contrasts even when the phylogeny is not fully resolved. Purvis et al. (1994) have shown that Pagel's (1992) method can be valid even when the topology of a phylogenetic tree and the branch lengths are not known in full.

(d) Models of evolution

The methods described in this section assign values at the internal nodes of the phylogeny, making assumptions about the way evolution is thought to have proceeded. One common assumption under the null hypothesis is that characters follow a Brownian motion pattern of evolution. Under this model, the evolutionary change in a character is equally likely to be in either direction and is independent of the value of the trait, when the null hypothesis is true. For example, body size could increase or decrease with equal probability at every time period along a branch of the phylogeny. If organisms did evolve in this way, then the possible evolutionary changes in a character along a branch of the phylogeny would have a mean of zero and a variance which is directly related to length of branch (assuming that branch lengths reflect time).

It is certainly inaccurate to assume that many features of organisms really evolve in this way, although for some characters it seems like a reasonable first guess. It is impossible to devise a method of calculating independent evolutionary contrasts on a phylogenetic tree that could account for the many different ways in which characters like geographical range size might evolve. In fact, two recent simulation studies (Martins and Garland, 1991; Purvis et al., 1994) suggest that the independent contrast

250 Predicting and understanding rarity

method is reasonably robust in the face of data generated according to models of evolution different from the one they assume, although some extreme results can be produced.

An alternative approach is to analyse the same data more than once, using different models of evolution, and to compare the results. Letcher and Harvey (1994) used a Brownian motion model in their study of geographical range sizes but considered it prudent also to use a model that estimated the range size at the internal nodes of two species as the sum of their current ranges, subtracting any overlap. They took these two versions of the independent contrast technique as being extreme cases and were 'reasonably confident of results that hold for both methods'.

13.4.3 The method of pairwise comparisons

Another, more radical, approach to the problematical issue of internal nodes is to confine analyses to comparisons between pairs of extant species, such that no two pairs share a branch of the phylogeny. For example, one might compare two finches, two ducks and two herons, with one rare and one common species in each pair. None of these pairs of species is linked together by a branch of the phylogeny shared by any other. No comparison is ever made between pairs of internal nodes. Thus, comparisons are independent and the method does not depend strongly on an explicit model of evolution if just the direction of the difference between each pair is used for statistical analysis. This sort of analysis is often referred to as 'sister taxa' comparisons, if the pairs are always two species that share an immediate common ancestor. It was suggested by Felsenstein (1985) and later developed by Burt (1989). Figure 13.8 shows the method more generally.

This approach is attractive by virtue of its simplicity but a weakness of the method is that the phylogeny does not uniquely define the pairs of comparisons, leaving them to the discretion of the investigator (Figure 13.8). The method also ignores much of the variation in the data, which may not always be the best way to proceed (Pagel, 1993: 201). Nevertheless, the method can be useful, especially when phylogenetic information beyond that of sister-species pairs is uncertain.

(a) Pairwise comparisons illustrated

Recall that in our initial analysis of the Australian birds, it was not possible to tell whether generalism or carnivory was the important characteristic that was related to a large geographical range size. We now classify species as being 'carnivorous' or 'herbivorous' independently of whether they are generalists or specialists.

We regressed across species the number of animal food types known to be eaten by each species against two independent variables in a multiple

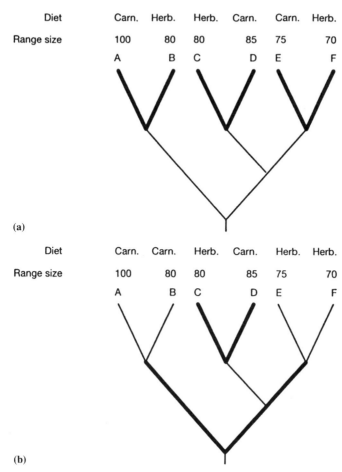

Diet	Carn.	Herb.	Herb.	Carn.	Carn.	Herb.
Range size	100	80	80	85	75	70
	A	B	C	D	E	F

(a)

Diet	Carn.	Carn.	Herb.	Carn.	Herb.	Herb.
Range size	100	80	80	85	75	70
	A	B	C	D	E	F

(b)

Figure 13.8 The method of sister taxa comparisons does not assume any model of evolution. In each of these two examples, we make comparisons defined by the thicker lines on the phylogenies. (a) We compare species A with species B and find that the carnivore has a larger range than the herbivore; the same is true when we compare species C with species D and again when we compare E with F. There are thus three independent occurrences of an evolutionary increase in range size being associated with a switch to a carnivorous diet, or vice versa. This set of comparisons is not unique – we could have compared species D with species F, although this would have precluded us from using any information about species C and E, since the comparison would not have been evolutionarily independent. In such a situation, it is sensible to make the maximum number of comparisons possible. (b) In this second example, however, it would be pointless to make a comparison between species A and species B because they do not differ in diet. Hence we compare A and B together on the one side with E and F on the other. We could do this by averaging the ranges sizes of the two groups, or by randomly choosing one species from each (e.g. comparing A with E, or B with F). In this example, there are only two independent comparisons where there is variance in both the variables of interest. Note that in this method, we can proceed without estimating what the common ancestor of A and B was like.

regression: the total number of food types that the species is known to eat, and the number of studies made on the species' diet. Residuals from this regression give a measure of the number of animal foods each species eats that is not confounded by the total number of food types it eats, nor by how well studied a species happens to be. Note that no evolutionary significance can be assessed from this across-species regression. For the purposes of this illustration the upper third of the species' residual values are taken as indicative of a strongly carnivorous habit, and the lower third as indicative of herbivory. We ignored the species falling in the middle third of residual values.

After classifying the species depending upon the value of their residual, we compared the range sizes of pairs consisting of one carnivore and one herbivore. The particular set of comparisons that can be made is not uniquely defined by the phylogeny (Figure 13.8). However, comparisons must be evolutionarily independent and once a given branch of the phylogeny has been considered, it cannot be included in a second comparison. Thus, it is simplest to choose pairs so as to allow the maximum number of comparisons possible with the data.

If there were no evolutionary association between diet and range size, we should expect half of the comparisons to show the carnivore having a larger range and half of them to show the opposite. In fact, out of 37 comparisons between carnivorous and herbivorous Australian birds, there are 32 in which the carnivorous species is more widely distributed across Australia. This is a very strong bias in one direction (binomial $P < 0.0001$) and we can conclude that, as a general rule, lineages of birds that have become carnivorous have also evolved a large range size, or that lineages becoming herbivorous have evolved a small range size.

13.4.4 Analyses within taxa

Sometimes, the available phylogenetic or taxonomic information is sufficiently untrustworthy that investigators wish to conduct their analyses in a way that does not depend upon specifying or even being able to identify the higher level (i.e. older) phylogenetic relationships among various groups. These are referred to these as 'within-taxa' methods (see also Pagel and Harvey, 1988) because analyses are confined to estimating some statistic of interest separately within a number of distinct groups.

For instance, Nee et al. (1991) calculated the relationship across species between body mass and population abundance in Britain, separately within 25 tribes of British birds. They found a positive relationship between the two variables in 18 of these tribes and concluded that, among closely related species, there was a significant trend for the largest species to be the most abundant. The important point from our perspective is that even though the separate tribal correlations were calculated across species, they were not separately tested for significance. Instead, these authors paid attention only to the pattern of slopes observed across all 25 tribes.

Interestingly, the same result can be achieved using a taxonomy and data on abundance from the eighteenth century. The Welsh naturalist Thomas Pennant (1768) grouped British bird species into genera (roughly the equivalent of what are now called tribes) and gave a qualitative impression of their abundances, based on information from a network of correspondents. For example, he says that the quail (*Coturnix coturnix*) was found throughout Britain 'but not in any quantity', and he attributes some of his information on this species to an anonymous 'gentleman, to whom this work lies under great obligations for his frequent assistance'. Using these 200-year-old data, there are 13 genera in which the species showed a positive size–abundance relationship and four genera with a negative relationship (binomial test $P = 0.05$, Figure 13.9). This is exactly the same pattern found by Nee and his co-workers, based on modern abundance data and a molecular phylogeny. Pennant's taxonomy would now be seen as hopelessly inaccurate, even in some cases giving separate species status to adults and juveniles of the same species.

Part of the reason why this result is repeated in such different data sets is that this particular method actually requires very little phylogenetic information. If we are satisfied that the groups we use are monophyletic, we can completely ignore the higher level phylogeny that interrelates these groups and also the relationships within each group. So, in Nee *et al.*'s (1991) and our analyses, it matters that the plovers and finches each form an independent tribe but the relationship between the finches and plovers is irrelevant. This was not true of the independent contrasts method used above to analyse correlates of range size in Australian birds, where contrasts were calculated across each node of the phylogeny. Even pairwise comparisons may require phylogenetic information, as in the second example in Figure 13.8.

Within-taxa analyses are a useful antidote to insufficient phylogenetic information but it would be unhelpful to think of them as a replacement for such information. When sufficient phylogenetic information is available to construct independent contrasts – including many instances in which the phylogeny is not fully resolved – this method is expected to provide a more powerful test of the hypothesis, by virtue of making use of more of the relevant information.

13.5 METHODS FOR THE ANALYSIS OF DISCRETE CHARACTERS

Comparative methods for the analysis of discrete characters make use of all of the same logic as those for continuous characters. However, because the variables take discrete values, calculating contrasts as we illustrated for continuous variables is not possible. Nonetheless, one could reconstruct,

Sign of mass–abundance relationship	18th century genus	Equivalent modern tribe	Sign of mass–abundance relationship
		Ducks	+ve
		Crows	+ve
		Flycatchers	+ve
		Chats	+ve
		Thrushes	+ve
		Wagtails	+ve
		Sparrows	+ve
		Tits	+ve
		Leaf warblers	+ve
		Blackcap,	
		whitethroats	+ve
−ve	Pheasants	Pheasants	+ve
+ve	Plovers	Plovers	+ve
+ve	Sandpipers	Sandpipers	+ve
−ve	Herons	Herons	+ve
+ve	Pigeons	Pigeons	+ve
+ve	Rails	Rails	+ve
+ve	Larks	Larks	+ve
+ve	Falcons and hawks	Falcons	+ve
		Hawks	−ve
−ve	Finches	Finches	−ve
+ve	Owls	Owls	−ve
+ve	Woodpeckers	Woodpeckers	−ve
+ve	Snipes, woodcock	Snipes, woodcock	−ve
+ve	Grebes	Grebes	−ve
+ve	Shrikes		
−ve	Gulls		
+ve	Terns		
+ve	Curlews		

using parsimony or other rules, the values at the internal nodes of the phylogeny for each of, say, two dichotomous variables. Then one could compare pairs of species and pairs of internal nodes of the phylogeny to see if, when one of the variables changes, the other changes in a systematic direction. This is a very simple 'independent contrast-like' technique for discrete characters, although it potentially ignores much of the variation in the data.

Other tests for discrete characters are those developed by Ridley (1983) and Maddison (1990). Ridley's test counts inferred changes in one or both of the characters throughout a phylogeny and asks whether changes in one character tend to be associated with changes in the other. However, the method counts only those branches in which a change occurs, and may therefore not be sensitive when changes in one or both characters are rare. Maddison's technique, on the other hand, makes use of all of the branches of the phylogeny and assesses whether an inferred pattern of changes in the two variables is unlikely, given all possible ways they could have evolved on the tree. This method is undoubtedly more powerful statistically than Ridley's.

However, neither of these methods for categorical variables makes use of branch length information, and they depend upon inferring a single set of values at the internal nodes of the phylogenetic tree (Pagel, 1994). Any inferences about the relationship between two characters will depend upon the validity of the unique set of reconstructed values. It is unusual in statistics that values inferred but not directly observed are nonetheless treated as actual observed data points.

Pagel (1994) has developed a method for the analysis of binary discrete characters that avoids the problem of inferring the pattern of changes at the internal nodes. The method finds evidence for correlated evolutionary change in two discrete characters by considering all possible transitions among states at each internal node. This avoids the logical problem of treating inferred values as actual observed data points, and is especially valuable for evolutionarily labile traits, such as many behavioural characters, for which no single most parsimonious reconstruction of the states at the internal nodes of the tree will exist. The method can also make use of information on branch lengths and can test hypotheses about the direction of evolutionary change.

Figure 13.9 In tribes of British birds, Nee *et al.* (1991) found that the largest are generally the commonest (right-hand side; we have re-analysed the data accounting for alterations in the molecular phylogeny). The same result could have been achieved in 1768 using the taxonomy of Thomas Pennant and abundance data gathered by him in correspondence with friends and colleagues. (Engraving of Pennant by J. Chapman (1823); those of birds are taken from Pennant's own book.)

The test begins by fitting two alternative statistical models to a data set. The model of 'independence' allows the two binary variables to evolve independently of one another, along each of the branches of the tree. Figure 13.10 shows that this requires four parameters. The model of 'dependent change' makes the probability of change in one variable dependent upon the state of the other variable. For example, one might be interested in whether the dependent variable Y is more likely to change from state 0 to state 1 when the independent variable is in state 0 versus in state 1. This model depends upon eight unique parameters (Figure 13.10).

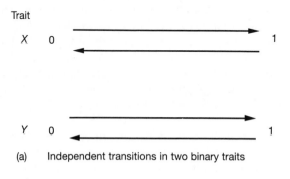

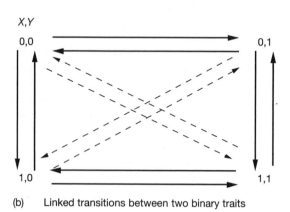

Figure 13.10 (a) Variables X and Y refer to two dichotomous traits measured on each of a number of species. Trait X can change from 0 to 1 or from 1 to 0, as can trait Y. If they are changing independently of one another, then the probability of a transition in X does not depend upon the state of Y in that species. (b) When changes in X or Y depend upon the state of the other variable, eight unique parameters are required to specify changes in one variable given the state of the other. The dashed lines indicate an instantaneous change in both variables and are not considered in the model.

The independent and dependent models of change are fitted to the data by maximum likelihood methods, which choose that set of values for the parameters of both models that make the observed data most likely, given the phylogenetic relationships among the species. This can determine whether the model of dependent change fits the data better than the model of independent change sufficiently well that we can disregard the latter as a useful description of the data.

To illustrate the method we have chosen two discrete characters from our data set. One is whether the species nests on the ground or not, the other is its IUCN (1986) status as 'threatened' or 'not threatened'. ('Threatened' is a category encompassing endangered, vulnerable and rare.) Each of the 80 species with the smallest geographical range sizes was so categorized and these data, along with their phylogenetic relationships, were analysed using Pagel's (1994) approach. We limited our analysis to the 80 least widespread species to identify the characteristics of rare species that are likely to make them threatened – not all species that are rare by any one definition are necessarily threatened with extinction.

The model of independence for these data has an overall log-likelihood of −95.54. Logarithms are used to simplify calculations, but do not change the results. The model of dependent change has a log-likelihood of −91.77. Because these values are logarithms of probabilities, more strongly negative numbers represent probabilities closer to zero (the logarithm of zero is negative infinity), whereas the less strongly negative number is the logarithm of a number closer to one (the logarithm of one is zero). Therefore we have some initial evidence that the model of dependent change is more probable. To assess whether this difference is significant, one uses the likelihood ratio statistic, defined as $LR = -2\log[I/D] = -2[\log(I) - \log(D)]$ where I and D stand, respectively, for the models of independence and dependence.

For the threatened/ground-nesting analysis, $LR = 7.54$. In this instance, the null hypothesis distribution of LR must be assessed by Monte Carlo simulation (Pagel, 1994). Figure 13.11 shows the results of a Monte Carlo simulation of the LR statistic for this set of data and phylogeny. Of the simulated values, 92% fall below the obtained value of 7.54 found for these data, and thus $P = 0.08$ for the relationship observed between nesting habit and IUCN threatened status, with ground-nesters more likely to be threatened.

One of the strengths of this method is that it allows us to go further and to test explicit directional hypotheses. In other words, we can go a further step down the road of distinguishing cause and effect. Figure 13.11 shows the values we obtained for the eight transition parameters of the model of dependence, each one corresponding to one of the arrows in Figure 13.10. We asked whether there is evidence that transitions from non-threatened to threatened status are more likely for ground-nesters than for tree-nesting species. Fitting a model to the data in which these two probabilities

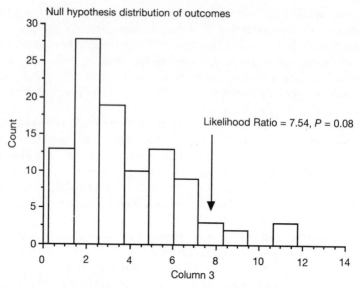

(a) Results of overall test: Nesting Type versus Threatened Status

L(I) = −95.54   L(D) = −91.77
LR = −2*[L(I) − L(D)] = 7.54, P = 0.08

(b) Individual transition parameters

1. Tree/Not Threatened to Tree/Threatened =	0.010904
2. Tree/Threatened to Tree Not Threatened =	0.204082
3. Tree/Not Threatened to Ground/Not Threatened =	0.058712
4. Ground/Not Threatened to Tree/Not Threatened =	0.000006
5. Tree/Threatened to Ground/Threatened =	0.048369
6. Ground/Threatened to Tree/Threatened =	0.150484
7. Ground/Not Threatened to Ground/Threatened =	0.246957
8. Ground/Threatened to Ground/Not Threatened =	0.224108

LR test of 1 versus 7 = 17.14, P < 0.00001

Figure 13.11 Top: Histogram of the results of a Monte Carlo simulation of the models of independence and dependence to test the likelihood ratio (*LR*) statistic. The simulations reveal that an *LR* value as extreme as 7.54 occurred eight times in 100 by chance, therefore *P* = 0.08 for this set of simulations. *Bottom*: The values for each of the eight transition parameters from the model of dependence (Figure 13.10) and the transition to which they correspond. The hypothesis of particular interest is whether being a ground-nesting bird puts that species at higher risk of becoming threatened than being a tree-nester. By forcing these two parameters to be equal in the model of dependence, a reduced seven-parameter model results.

are forced to be the same (although the precise value depends on the real data) and comparing it with the model of dependence reported above (in which the two probabilities are allowed to vary) yields a LR statistic of 17.14. In this instance, LR is distributed as a chi-squared variate with one degree of freedom, and 17.14 is significant at $P < 0.00001$: ground-nesting species are much more likely to become threatened than are tree-nesting species. We can be certain in this instance that ground-nesting is not a consequence of rarity – it happens first. It is still possible that ground-nesting is not the cause of rarity; a third, unidentified variable that is correlated with ground-nesting could be involved.

13.6 DISCUSSION AND SUMMARY

We have described several different comparative methods that can be used to study rarity and its correlates. One useful way to classify these methods is along an axis of varying amounts of phylogenetic information. At one end are the cross-species analyses that require no phylogenetic information but which cannot be relied upon to yield trustworthy results, if one wants to know about the evolutionary correlates of rarity. By this we mean: what are the evolutionary changes or adaptations that ironically place a species at higher risk of becoming rare in the future? At the other end of the continuum are the techniques of independent contrasts for continuous variables and the methods for discrete characters. These methods will reliably identify evolutionary correlations, and in some cases explicit directional hypotheses can be tested.

In the middle of the continuum are the paired analyses and the within-taxa analyses. These techniques are useful either when one does not want to rely on inferring values at the internal nodes (pairwise) or when the phylogenetic information at the higher levels of the phylogeny is untrustworthy (within-taxa methods). However, one should bear in mind that both the pairwise and the within-taxa methods ignore much of the variation in the data. Also, in the case of pairwise methods, the phylogeny does not uniquely define which species should be paired.

Using the results from the various methods, we can now draw a profile of a bird species likely to be rare in Australia. It will belong to a phylogenetically old taxon and live in a scrub habitat; it will be a ground-nester with a small clutch and have a narrow, plant-based diet. Given that such species are expected to be under threat, it is perhaps not surprising

---

This model had a log-likelihood of -100.34 versus -91.77 for the full eight-parameter model. The likelihood ratio statistic is therefore 17.14, which as a chi-squared variate with one degree of freedom is highly significant: ground-nesting birds are at significantly higher risk of becoming threatened than are tree-nesting birds. Note that the individual transition parameters are not P-values associated with any statistical null hypothesis.

that none exists among the Australian avifauna, although the ground parrot (*Pezoporus wallicus*) comes close. The parrots are one of the oldest tribes and this species does indeed nest on the ground in scrubland and has a relatively restricted vegetarian diet. Its clutch of three to four eggs is about the average for parrots and is thus small but not especially so (more than half of the species in the study have four or fewer eggs).

With all the worst attributes it could have, the species is indeed rare. It has probably never been common but the subspecies *wallicus* is now extinct in South Australia and nearly so in Queensland. It is suffering from habitat destruction and is scarce in almost all of what remains of its range (King, 1981; Collar and Andrew, 1988). In 1986, it was considered rare (not endangered) but now its survival is considered unlikely (IUCN, 1986; Groombridge, 1994). If only it had known about the comparative method, it could have recognized the value of eating meat or nesting in trees, and saved itself.

REFERENCES

Anon. (1990) *Reader's Digest Complete Book of Australian Birds,* Reader's Digest, Sydney.

Barker, R.D. and Vestjens, W.J.M. (1990) *The Food of Australian Birds*, Vols 1–2, CSIRO, Melbourne.

Blakers, M., Davies, S.J.J.F. and Reilly, P.N. (1984) *The Atlas of Australian Birds*, Melbourne University Press, Melbourne.

Burt, A. (1989) Comparative methods using phylogenetically independent contrasts. *Oxford Surveys in Evolutionary Biology*, **6**, 33–53.

Collar, N.J. and Andrew, P. (1988) *Birds to Watch: The ICBP World Checklist of Threatened Birds,* ICBP, Cambridge.

Cotgreave, P. (1994) Patterns in species abundances. *Science Progress*, **77**, 57–69.

Cotgreave, P. (1995) Population density, body mass and niche overlap in Australian birds. *Functional Ecology*, **9**, 285–289.

Cotgreave, P. and Harvey, P.H. (1991) Bird community structure. *Nature*, **353**, 123.

Cotgreave, P. and Harvey, P.H. (1992) Relationships between body size, abundance and phylogeny in bird communities. *Functional Ecology*, **6**, 248–256.

Cotgreave, P. and Harvey, P.H. (1994) Biogeographic and phylogenetic association with bird species diversity. *Biodiversity Letters*, **2**, 46–55.

Felsenstein, J. (1985) Phylogenies and the comparative method. *American Naturalist*, **125**, 1–15.

Grafen, A. (1989) The phylogenetic regression. *Philosophical Transactions of the Royal Society of London, B*, **326**, 119–156.

Grafen, A. (1992) The uniqueness of the phylogenetic regression. *Journal of Theoretical Biology*, **156**, 405–423.

Groombridge, B. (1994) *1994 IUCN Red List of Threatened Animals,* IUCN, Cambridge.

Harvey, P.H. and Pagel, M.D. (1991) *The Comparative Method in Evolutionary Biology*, Oxford University Press, Oxford.

IUCN (1986) *The 1986 IUCN Red List of Threatened Mammals*, International Union for the Conservation of Nature and Natural Resources, Gland, Switzerland.

King, W.B. (1981) *Endangered Birds of the World: The ICBP Bird Red Data Book*, Smithsonian Institution Press, Washington, DC.

Lawton, J.H. (1993) Range, population abundance and conservation. *Trends in Ecology and Evolution*, **8**, 409–413.

Letcher, A.J. and Harvey, P.H. (1994) Variation in the geographical range size among mammals of the Palearctic. *American Naturalist*, **144**, 30–42.

Mace, G.M. (1995) Classification of threatened species and its role in conservation planning in *Extinction rates* (eds J.H. Lawton and R.M. May), Oxford University Press, Oxford, pp. 197–213.

Maddison, W.P. (1990) A method for testing the correlated evolution of two binary characters: are gains and losses concentrated on certain branches of a phylogenetic tree? *Evolution*, **44**, 539–557.

Martins, E.P. and Garland, T.H. (1991) Phylogenetic analyses of the correlated evolution of continuous characters: a simulation study. *Evolution*, **45**, 534–557.

Mooers, A.O. and Cotgreave, P. (1994) Sibley and Ahlquist's tapestry dusted off. *Trends in Ecology and Evolution*, **9**, 458–459.

Nee, S., Read, A.F., Greenwood, J.J.D. and Harvey, P.H. (1991) The relationship between abundance and body size in British birds. *Nature*, **351**, 312–313.

Pagel, M. (1992) A method for the analysis of comparative data. *Journal of Theoretical Biology*, **156**, 431–442.

Pagel, M. (1993) Seeking the evolutionary regression coefficient: an analysis of what comparative methods measure. *Journal of Theoretical Biology*, **164**, 191–205.

Pagel, M. (1994) Detecting correlated evolution on phylogenies: a general method for the comparative analysis of discrete characters. *Proceedings of the Royal Society of London, Series B*, **255**, 37–45.

Pagel, M. and Harvey, P.H. (1988) Recent developments in the analysis of comparative data. *Quarterly Review of Biology*, **63**, 413–440.

Pennant, T. (1768) *British Wildlife*, Vols. 1–3, Benjamin White, London.

Purvis, A. (1992) Comparative methods: theory and practice. DPhil. Thesis, University of Oxford.

Purvis, A., Gittleman, J.L. and Luh, H.-K. (1994) Truth or consequences: effects of phylogenetic accuracy on two comparative methods. *Journal of Theoretical Biology*, **167**, 293–300.

Ridley, M. (1983) *The Explanation of Organic Diversity: the comparative method and adaptations for mating*, Oxford University Press, Oxford.

Sibley, C.G. and Ahlquist, J.E. (1990) *Phylogeny and Classification of Birds: a study in molecular evolution*, Yale University Press, New Haven.

14 Concluding comments

Kevin J. Gaston and William E. Kunin

14.1 INTRODUCTION

As the preceding chapters demonstrate, considerations of rare–common differences ramify in many directions. The exploration of possible mechanisms touches on fundamental issues in ecology, evolution and paleontology, to name but a few. In this final chapter we draw together some broad messages from this volume, and identify some important lacunae in coverage of the topic of rare–common differences. In doing so, we hope to stimulate further work in what we find an exciting field of research.

14.2 ESTABLISHING PATTERNS

The working premise of this volume is that there are some repeatable, and perhaps quite general, differences in the average biological traits exhibited by rare and common species (reviewed in Chapter 2). We believe that this is indeed so. Nonetheless, there are some significant caveats. First, there must be some doubts as to how good an insight the published literature provides into real patterns of rare–common differences. There are several reasons for this:

- Studies which document broad simple rare–common differences are more likely to be written up and published than are those which find no relationships or very complex interactions (the 'file drawer problem').
- Reliable data on the biologies of very rare species are often not available, and these are therefore not included in some analyses (Chapters 7 and 13).
- The evaluation of some biological traits of species is strongly dependent on sample size (e.g. environmental tolerances, habitat breadths, clutch size variation, longevity), making it difficult to discern whether apparent differences between rare and common species are genuine or are direct consequences of differences in sample sizes associated with differences

The Biology of Rarity. Edited by William E. Kunin and Kevin J. Gaston.
Published in 1997 by Chapman & Hall, London. ISBN 0 412 63380 9.

in the abundances or distributions of species (Gaston, 1994). Indeed, the measurement of abundance itself may be complicated by sample-size related problems (Chapter 4).

These complications may result in a somewhat biased view of the nature and causes of rare–common differences, but they seem very unlikely to explain the patterns observed.

Second, the diversity of ways in which studies have been performed greatly impacts on the interpretation and comparison of analyses of rare–common differences. This variation may embrace:

- the definitions and measures of rarity applied (Chapter 3; Gaston, 1994);
- the definitions and measures of other biological traits;
- the spatial scale;
- the numbers of species;
- whether phylogenetic effects are considered.

Such differences may themselves reflect differences in the objectives of studies, in the ways in which analyses are routinely performed for a given taxon or in a given area, as well as the practical limitations posed by the availability and ease of acquisition of data. Although from some perspectives this variation can be regarded as an obstacle to our understanding of rare–common differences, it can also be viewed as providing a further justification for arguing that the patterns documented are indeed very general.

The recognition both of potential biases in documented rare–common differences and of factors which complicate their interpretation and comparison is arguably a constructive exercise. It highlights the need for caution in carrying out analyses of such differences, particularly with regard to factors which may generate artefactual patterns. It also reveals the possibly great benefits which would be obtained from collecting and analysing data for a variety of assemblages in a similar manner, and from analysing data for a given assemblage in a variety of ways. With respect to the latter, there seems to be a great need for combining detailed experimental and observational work on rare–common differences between a few species or groups of species belonging to a larger species assemblage, with the exploration of broad statistical patterns of rare–common differences across all species in the same assemblage.

14.3 BROAD TRENDS AND PREDICTIVE POWER

Accepting that there is evidence for some general, repeatable differences between the biologies of rare and common species, we must be clear what we mean by this. These are broad patterns. They do not necessarily have great power to predict reliably the traits that will be exhibited by any given individual species; the variance about patterns of rare–common differences may be quite large. Nor can they be regarded as identifying a discrete set of rare species characterized by a suite of consistent traits. Rather, they predict average differences between the biological traits exhibited by rare and common species in an assemblage. To say that rare species on average have greater levels of, for example, homozygosity than common species (e.g. Chapters 2, 10) is not the same as saying that the abundance or range size of a species is a direct indication of its level of homozygosity (see McLaughlin, 1992, for an analogous argument concerning the relation between the age of a species and the size of its geographical range).

14.4 DIFFERENTIATING MECHANISMS

One of the most important points of this book is that a wide variety of different mechanisms can account for very similar patterns. If rare and common species (however defined) express rather different suites of traits, those differences may be due to biases in community assembly rules (Chapter 5), evolved responses to rarity (Chapters 11, 12), genetic (Chapter 10) or even plastic (Chapter 9) responses to abundance, or differences in speciation and/or extinction rates among different groups (Chapters 6, 7, 8). These various mechanisms are not, for the most part, mutually exclusive, and so any given pattern may be due to more than one process, further muddying the waters. How can we possibly hope to sort through such a shopping list of options to find out which are most important?

It may help somewhat to divide the various options into the three categories we have used to organize this volume: entry rules, exit rules and transformations (Chapter 1). **Entry rules** suggest that species with certain traits are more likely to become rare in the first place. **Exit rules** edit the set of rare species, suggesting that species with certain traits are likely to persist longer once they have become rare. **Transformations** take place when the traits in question develop as a result of rarity. In a sense these three categories can be further reduced to two. In both entry and exit rules, possession of certain traits **causes** (or, in the case of exit rules, **allows**) rarity, whilst in the case of transformations the trait develops as a **consequence** of rarity. The fundamental question thus comes down to which came first, rarity or the traits found to be associated with it? And how could we tell the difference?

Ideally, we would like to have a fleshed-out history of the characteristics and abundance (or range size) of each species from the moment it differentiated until its extinction. Given such a record, it would be a simple matter to note just how often rarity follows the acquisition of a given trait, and how often traits are acquired following the onset of rarity. Indeed, such a history would tell us a great deal about the dynamics of rarity over a species' career (Figure 14.1). An implicit assumption in many analyses is that a species' abundance is essentially fixed: a species quickly achieves its destined level of commonness or rarity and maintains it until it crashes to extinction (Figure 14.1a). Alternatively, we might assume (with Willis, 1922) that most species are initially rare and restricted, but gradually expand their empires over time (Figure 14.1b). It is certainly true that many modes of speciation require species to begin their lives with very small numbers and very limited ranges, and the phenomenon of 'neo-endemism' is well documented. The opposite pattern also has some support: species may decline gradually in abundance over time, leaving very old taxa as rare, local 'paleoendemics' in isolated scraps of refugial habitat (Figure 14.1c). These various perspectives could also be combined, with species gradually rising to some peak of abundance, and then sinking gradually back towards extinction (Figure 14.1d). Finally, even more complex dynamics are possible, in which the abundances of species fluctuate widely and erratically, with multiple plateaus or none at all (Figure 14.1e). Knowing something of the place of rarity in the career of a species would tell us a great deal about the likely causes of rare–common differences. Evolutionary adaptations to rarity, for instance, seem most

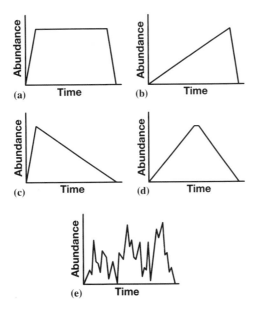

Figure 14.1 Examples of how the abundance of a species might change over its lifetime. (a) A species may attain a relatively constant abundance and maintain it for much of its career. Some suggest that (b) species abundance tends to rise gradually over time, or (c) decline progressively, or (d) rise initially and then decline. Alternatively, (e) abundance might shift dramatically and rather unpredictably over time. These different trajectories would have different implications for the relative importance of entry biases, exit biases and transformations in moulding the set of rare species.

plausible if species tend to remain rare for evolutionarily significant periods of time, ideally over much or all of their history.

Unfortunately, the imperfections of the fossil record (Chapter 7) suggest we will never be able to compile such a fully fleshed-out history, certainly not for more than a few species. Is there anything that we can do with the information we do have? One promising possibility is to look into the phylogenetic relationships among species (Chapter 13). As the quality and extent of phylogenetic information improves, it may provide us with strong (albeit indirect) evidence about the timing of trait acquisition. If, for example, a rare wingless species of beetle was a relatively recent offshoot of a clade in which older and commoner species are also wingless, it would be hard to argue that winglessness arose as an adaptation to the conditions of rarity. If most rare species tended to appear within such generally wingless clades, it would be reasonable to conclude that rarity was more likely to be a result of winglessness (due to some entry or exit rule) than a cause of it (through some transition rule) in this group (Figure 14.2a). Wingless beetle clades might be more likely to produce rare species by local speciation events due to reduced gene flow, or perhaps rare local populations can better persist in wingless groups, because the drain of emigration on local population dynamics is reduced. Conversely, if rare species tended to be clustered in certain clades, and some trait (e.g. winglessness) arose disproportionately among some species of those clades (Figure 14.2b), a good case could be made for rarity leading to the acquisition of the trait in question. If, however, every wingless species was rare and every rare species was wingless (Figure 14.2c), it would be impossible to deduce whether rarity was the cause or the effect. Perversely, the better the correspondence between rarity and a given trait, the more difficult it is to draw conclusions about which was the cause and which the effect. Fortunately or unfortunately, given the generally indistinct nature of most recorded rare–common differences (section 14.2), this is unlikely to be a problem for us.

Phylogenetic records may, indeed, be able to tell us more than just the relative timing of trait acquisition. Under certain circumstances, good phylogenies may allow specific mechanisms to be identified for some rare–common differences. If a phylogeny shows clades bearing certain traits to be rapidly speciating, and that rarities tend to be clustered in such clades (Figure 14.2d), a mechanism of selective speciation seems to be implicated in establishing the rare–common difference. Phylogenies may also prove useful, with a bit of care, for identifying likely cases of selective extinction. Recent work by Nee *et al.* (1995) suggests that extinction rates within clades can be deduced from patterns of branching over time. Groups with high extinction rates should show a disproportionate flowering of species in the very recent past – shallow top branches on their phylogenetic trees. In well studied phylogenies, such excess numbers of recent lines represent species that have not had sufficient time to become extinct, and the greater the surplus, the larger the implied extinction rate. This suggests that a selective extinction model of rare–common differences might be detectable in a phylogenetic tree by comparing the age of rare taxa bearing different traits. If clades with certain traits (e.g. self-incompatible plants) have rare species only (or very disproportionately) on shallowly rooted top branches of a phylogenetic tree, while rare species with other traits (e.g. self-compatible species) are divided more evenly among deep and shallow branches, the implication is that the first trait increases extinction rates when rare, whereas the second allows for persistence (Figure 14.2e).

As yet, there is overwhelming evidence in favour of no single mechanism as a determinant of observed rare–common differences. Rather, all three (entry rules, exit rules, transformations) seem intuitively reasonable, and all receive some, albeit often indirect, empirical support in this book.

14.5 ABUNDANCE AND EVOLUTION

The ongoing controversy as to whether rarity increases or decreases rates of evolutionary change has been touched upon repeatedly in this volume (Chapters 5, 6, 7, 10, 12). If the issue remains unresolved, it nonetheless seems to be coming into clearer focus. To a large extent, the answer one gets seems to depend on the definitions employed; it seems that the effect of rarity on evolution differs depending on the form of rarity and the mode of evolution involved. In Chapter 12, a wide array of cases are considered, models in which the rate of evolutionary change within a population could be shown to increase, decrease or be unaffected by the population size or geographical range of a species. The nature of the effect seems to depend (among other factors) on whether a population is open to immigration or isolated from it, on whether the trait involved is neutral or selectively favoured, and on whether the adaptive landscape is stable or shifting. A similarly complicated picture is developing concerning the effects of abundance on rates of evolutionary diversification. One repeated observation is that traits which augment speciation rates also often augment extinction rates (Chapters 6, 7). This suggests that the set of rare species may turn over much more quickly than the set of common ones. It also implies that the net rate of diversification (speciation rate minus extinction rate) will be sensitive to the precise form of the relationships between abundance and each of the two component processes.

14.6 RARITY TRAPS

One other aspect of evolution in rare species is worth mentioning here. Natural selection generally favours traits that increase population size; it is hard to imagine a circumstance in which a species would evolve rarity. Nonetheless, it seems probable that a species, if it is kept rare by extrinsic factors over evolutionarily significant stretches of time, may evolve traits

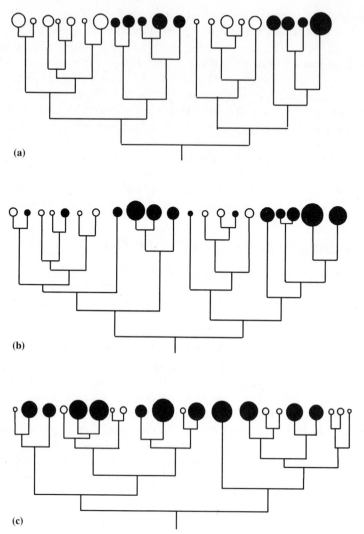

Figure 14.2 Hypothetical phylogenies demonstrating how the causes and consequences of rarity might be distinguished. Circles represent species, with size indicating abundance on some scale (large = common). *Open circles* possess some trait which is disproportionately found among rare species (e.g. winglessness or self-compatibility); *solid circles* have the opposite characteristic (e.g. wings or self-incompatibility). (a) Where rarity tends to occur primarily in clades having the associated trait, that trait seems to be implicated in causing (or allowing) rarity. (b) Conversely, if the trait in question occurs only in clades of rare species, rarity may be responsible for the trait's development. (c) Ironically, if the correspondence between abundance and the trait is absolute, these hypotheses cannot be differentiated. Data on the branch depths could allow greater resolution. (d) The trait in question may increase speciation rates and thereby include more recently created rarities. (e) Alternatively, the trait may allow the persistence of rare species over time, whereas rare species with other traits quickly go extinct.

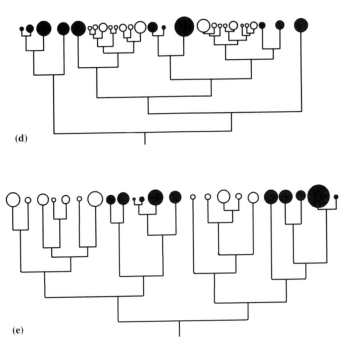

Figure 14.2 contd.

that adapt its members to the condition of rarity (e.g. Chapter 11). Some of these traits may, in turn, make it more difficult for the species ever to become common should the opportunity arise. Thus, we arrive at the notion of a 'rarity trap'. The trap involved is rather different from that invoked in Chapter 12 (also cf. Chapter 6), where it was suggested that a rare marginal 'sink' population of a species may be prevented from adapting to its local environment by a flow of immigrants from commoner 'source' populations. Instead, the rarity trap we propose here results precisely when a rare species **does** adapt to its situation.

The rarity trap, thus defined, is only a special case of a more general phenomenon, which might be termed the 'adaptation trap'. Many evolutionary adaptations which improve the ability of a species to exploit some specialized and under-utilized niche-space may preclude the species from taking advantage of new opportunities which may arise in the future. A short-term gain is bought at the cost of longer-term imprisonment. Adaptations are only a 'trap', of course, if they cannot be un-evolved when the new opportunities present themselves. There are at least four reasons why this might occur. The first has to do with asymmetries in rates of evolution: quite simply, it is easier to lose a structure than it is to reinvent it. An adaptation may also become fixed by subsequent evolution that relies upon its existence. Thus a trait which originally evolved as a foraging device may come to have a role in thermoregulation or sexual display, and thus may not be easily jettisoned should the original selective pressures which led to its evolution become reversed. Conversely, characters which are not used in the species' new specialized lifestyle (e.g. beetle wings or orchid nectaries) may be degraded by the evolutionary equivalent of a 'use it or lose it' rule. There will be little or no selection against mutations that cause an unused structure to degenerate, but the build-up of such mutations would make it difficult to return to using that structure in the future (D. Cohen, personal communication). Finally, ecology (as all of nature) abhors a vacuum. As a consequence, if and when new opportunities present themselves, they will be quickly taken up by some species less constrained by inconvenient 'excess baggage'.

A potential example of a rarity trap is provided by sparsely populated plant species, which may face difficulties in attracting pollinators (Chapter 9; Rathcke, 1983; Feinsinger et al., 1991; Kunin, 1993). One solution to this difficulty is to mimic the flowers of other commoner species (Dafni, 1984). This arrangement will be particularly suitable in groups that already have unusual mechanisms of pollen placement or packaging (e.g. orchids and ascleps), which can partially solve the problem of lost pollinator effectiveness that would otherwise result from mimicry. Once a mimetic pollination system is firmly in place there ceases to be any significant selective pressure to offer rewards to pollinators, and it is a very common finding that mimetic flowers offer none. Therein lies the rarity trap. Once a species has unrewarding flowers, it seems unlikely that it could ever again become common, even if conditions were to change in a way that would otherwise have permitted it. Without floral rewards, reproductive success for a mimic should decline rapidly as its population densities rise, as the ratio of mimics to models would shift inexorably. With mimics common, pollinators would have strong incentives to learn to differentiate the rewarding and unrewarding species or, failing that, to avoid the pair of them altogether. The resulting decline in pollination success could operate as a powerful feedback loop, maintaining the mimic's population at low density.

The possible existence of the rarity trap serves to reinforce the necessity of performing experiments to demonstrate why individual species are rare. Observation of particular attributes of rare species may suggest that their rarity results from pressures which do not in fact exist. This may be of importance when rapid assessments of the determinants of a species' rarity have to be made so that conservation action can be formulated.

14.7 IMPLICATIONS OF RARE–COMMON DIFFERENCES: CONSERVATION BIOLOGY

Rare species have been attracting increasing attention in recent years. This has been due not so much to a passion for the ideas that occupy this volume, but rather to an acute and probably overdue concern for their preservation. If rare and common species differ in biologically significant traits (and it appears that they do), this may have implications for the way such conservation efforts are performed.

One particularly striking case in point is the 'model species' approach used widely in conservation decision-making. Very little is known about the characteristics of most rare species. Consequently, management decisions must often be made without adequate information. It is a common practice to use data from better-studied common relatives of rare species where data from the species itself are lacking. This practice introduces a potential source of error: no two species are identical. But it may also introduce a bias: if rare and common species show different distributions of characters, the difference between a well known common species and its poorly studied rarer relative may not be a random error. On the face of it, this argues for further caution when using data from one species to plan management for another. Nonetheless, data from such model species are often the only data available, and so may be indispensable. Rather than disregarding them, we might learn to use them more wisely; by documenting patterns of rare–common differences, we may be able to correct such observations for bias. If, for example, a population viability analysis requires an estimate of the (unknown) reproductive rate of some rare rodent, we may be able to use data from better studied common congeners and correct the estimate downward to account for the generally lower reproductive effort found in rare species.

The above consideration holds regardless of the cause of the rare–common difference involved. Conservation biology might also profit from a greater understanding of the relative importance of the different mechanisms outlined in this volume. If, for instance, transforming processes (e.g. evolutionary adaptation) are responsible for many or most rare–common differences, then we might expect species recently made rare due to anthropogenic causes to have rather different suites of characters than those exhibited by species that have been rare over evolutionarily significant periods (Chapter 11). It is possible that species with long evolutionary experience of rarity may have adapted in ways that make them less vulnerable to extinction than they might otherwise have been. A similar situation might apply if selective extinction has removed rare species with traits that made them particularly extinction prone (Chapters 7, 8); again, newly rare species might be expected to be at greater risk than species with a history of rarity. Such differences, if they were confirmed, would suggest that management efforts should be concentrated on recent, anthropogenic rarities, which may be in greater need of our assistance. If the rarity trap (section 14.6) proves to be important, such newly rare species might not only be more vulnerable to the conditions of rarity; they may also be more responsive to management practices designed to ameliorate them.

14.8 CODA

To conclude on such an applied note seems odd, in a book largely devoted to rather theoretical issues about the origin and maintenance of pattern. Nonetheless, we believe it is appropriate, given the fact that perhaps the most important difference between rare and common species is in the threats they face at human hands. Collectively, we have brought about a spasm of extinctions which may well sweep away half of the species of life on this planet within a few decades (Ehrlich, 1995). A disproportionate share of that extinction burden will inevitably fall upon rare species, which by virtue of their small ranges and limited numbers are already close to the threshold. If the study of rarity can help us in some small way better to preserve the diversity of life, the pursuit will be of more than theoretical value. This most crucial difference between rare and common species is one which we may yet have the power to eliminate.

REFERENCES

Dafni, A. (1984) Mimicry and deception in pollination. *Annual Review of Ecology and Systematics*, **15**, 259–278.

Ehrlich, P.R. (1995) The scale of the human enterprise and biodiversity loss, in *Extinction Rates* (eds J.H. Lawton and R.M. May), Oxford University Press, Oxford, pp. 214–226.

Feinsinger, P., Tiebout, H.M., III and Young, B.E. (1991) Do tropical bird-pollinated plants exhibit density-dependent interactions? Field experiments. *Ecology*, **72**, 1953–1963.

Gaston, K.J. (1994) *Rarity*, Chapman & Hall, London.

Kunin, W.E. (1993) Sex and the single mustard: population density and pollinator behavior effects on seed-set. *Ecology*, **74**, 2145–2160.

McLaughlin, S.P. (1992) Are floristic areas hierarchically arranged? *Journal of Biogeography*, **19**, 21–32.

Nee, S., Holmes, E.C., May, R.M. and Harvey, P.H. (1995) Estimating extinction from molecular phylogenies, in *Extinction Rates* (eds J.H. Lawton and R.M. May), Oxford University Press, Oxford,, pp. 164–182.

Rathcke, B. (1983) Competition and facilitation among plants for pollination, in *Pollination Biology* (ed. L. Real), Academic Press, Inc., Orlando, FL, pp. 305–329.

Willis, J.C. (1922) *Age and Area: a study in geographical distribution and origin of species*, Cambridge University Press, Cambridge.

Index

DATE DUE

FEB 0 4 2002	
MAY 0 5 2010	
MAY 0 8 2012	
JUN 21 2012	